THE FIRST DOMESTICATION

350/450

"THE EYES OF A SLEEPING VILLAGE"

REALON LIMITED TRADER 8105

RAYMOND PIEROTTI

AND BRANDY R. FOGG

The First
Domestication

HOW WOLVES AND HUMANS COEVOLVED

Yale UNIVERSITY PRESS   NEW HAVEN AND LONDON

Yale University Press books may be purchased in quantity for educational, business, or promotional use. For information, please e-mail sales.press@yale.edu (U.S. office) or sales@yaleup.co.uk (U.K. office).

Set in Scala and Scala Sans type by IDS Infotech, Ltd., Chandigarh, India.
Printed in the United States of America.

Frontispiece: *Eyes of a Sleeping Village,* by Cheyenne artist Merlin Little Thunder (Courtesy of the artist)

Library of Congress Control Number: 2017936094
ISBN 978-0-300-22616-4 (hardcover : alk. paper)

A catalogue record for this book is available from the British Library.

This paper meets the requirements of ANSI/NISO Z39.48-1992 (Permanence of Paper).

10 9 8 7 6 5 4 3 2 1

To Cynthia Annett, who always provides wise guidance,
and to the memories of Tabananika, Nimma, and Peter:
companions, friends, and teachers

CONTENTS

When the first modern humans moved out of Africa and into the vast continent of Asia, wolves were the most widely distributed large mammal in the world, ranging across the entire Northern Hemisphere. To achieve this, they had to have been highly adaptable to conditions ranging from the subtropics of southern Asia to the Arctic regions of Asia and North America. There was no obvious reason for wolves to ally themselves with bipedal primates—and, in fact, many believe that they did not choose this association with early humans; it was forced upon them.

In our efforts to produce a significant new contribution to a crowded field, we looked to a source that has been largely ignored in investigations of the evolution of humans and their ecological relationships with other species: the solid information contained within accounts from Indigenous peoples around the world. When we began work on this manuscript, Brandy Fogg was a master's student in the Indigenous Nations Studies Program at the University of Kansas, and Raymond Pierotti was a faculty member with an appointment split between that program and the Department of Ecology and Evolutionary Biology at KU. Our goal was to examine the relationship that nondomestic wolves had with Indigenous peoples by studying how the Indigenous peoples themselves described and characterized that relationship.

Looking at the human/wolf relationship from this perspective is important because, with one possible exception (Rose 2000, 2011; see below), every other book that has examined the history of the domestication of wolves has been written primarily or exclusively from a contemporary Western (European or Euro-American) point of view. Many of these books are excellent and insightful, but authors from the Western tradition almost always assume that early relationships between humans and wolves involved competition, violence, and hostile rivalry rather than cooperation and compatibility. This perspective rests on either of two major assumptions: (1) that humans "stole" wolf puppies from dens, or (2) that some wolves, unable to hunt for themselves, took scraps from human refuse dumps and in the process evolved on their own into dogs. The overriding theme in both assumptions is that these wolves became dependent on humans, who controlled the relationship. Finally, it is assumed to have taken thousands of years of selective breeding to turn wolves into the domestic dogs we see today.

In marked contrast to this Western perspective are accounts from Indigenous peoples, who invariably describe congenial interactions between wolves and humans involving the freedom of wolves to come and go from human encampments as they chose. These accounts provide evidence that such relationships lasted well into recent times, ending only when the peoples themselves were forced off their lands and out of their traditional ways of life as the wolves simultaneously came under genocidal attack through the pressures of colonialism. In many cases, Indigenous peoples explicitly describe how their companions the wolves suffered similar or worse levels of persecution by invading Europeans. Indigenous narratives do not describe stealing puppies but playing with puppies. In a few instances, they report rescuing puppies from dens when the parents could not provide for their entire litters. The humans raised these puppies, some women actually nursing the puppies from their own breasts, and when the pups reached sexual maturity, they would leave the humans but return periodically to hunt with them or to cooperate in making it through a hard winter. Basically, both wolves and Indigenous peoples knew that their lives were longer and more secure when they shared them together.

A story recently told by a Native American scholar illustrates this type of relationship, although it describes a domesticated animal rather than a wild one:

One . . . traveler that crossed my path was a dog. Where he came
from we never knew, nor . . . where he went after his sojourn
with us. . . . One evening there was a loud commotion. . . . My
grandfather and I found the reason: a skinny, gray dog . . . stood
his ground . . . hardly annoyed by the snarling feints [of our
dogs]. . . . The commotion ceased and my grandfather tossed
him a scrap of meat. The greyhound delicately picked it up and
trotted off. . . . I thought he was gone for good. . . . But I was
wrong. The next morning he was sitting . . . waiting. . . . I chose
to ignore him. My grandmother . . . fed him. I guessed that he
decided to stay around for those scraps, until I saw him catch a
rabbit. My grandfather and I . . . noticed . . . the greyhound . . .
following us . . . when a jackrabbit exploded from a low thicket
and bounded across the prairie. The greyhound flashed past. . . .
The rabbit accelerated, moving faster than I had ever seen any
living thing move. It seemed to glide above the grass. Foolish
dog, I thought . . . you will not catch that rabbit. . . . As the rabbit
hit top speed, the dog seemed to angle to the left and . . . was a
grey blur through the grass. . . . Jackrabbits . . . have a peculiar
habit of making a wide circle when pursued. No four legged
hunter on the prairie can match their straightaway speed. . . .
They eventually make a wide turn and end up near . . . their den
or burrow, having easily outrun any pursuit. . . . Off in the
distance, perhaps a hundred yards, two gray forms collided,
then came a faint squeal, then silence. . . . The greyhound . . .
hit the jackrabbit in mid-stride, closing on him at an angle . . .
[and] caught himself a sumptuous meal. . . . I couldn't believe
the dog had outsmarted the jackrabbit. . . . He shared his food
with our dogs, after dragging the rabbit's carcass across the
prairie. . . . He stayed with us over the winter until early spring.
Then, just as suddenly as he had appeared, he took his leave.
(Marshall 2005, 41–44)

The author of this account, Lakota scholar Joseph Marshall III, has been
one of our inspirations. His book *On Behalf of the Wolf and the First Peoples*
(1995) provided us with a basic framework for some of our arguments as
well as confirmation that we were on the right track.

The approach we take was inspired by the work of the insightful German ethologist Wolfgang Schleidt, who in 1998 posed the question "Is humaneness canine?" Schleidt argued that much of the behavior of modern humans can be traced to their relationship with wolves and dogs. Schleidt followed this up with a more developed argument that posited an "alternative view of dog domestication," suggesting that the scientific name for modern humans might be better *Homo homini lupus* (Schleidt and Shalter 2003). We have followed the path first blazed by Schleidt, combining his ideas with our knowledge and experience of Indigenous peoples and their traditions.

The extensive scholarship about dogs and domestication has been well reviewed by the creative and clever archaeologist Darcy Morey, one of the world's experts on burials of dogs by humans and the author of a fine book on the human/dog relationship, *Dogs: Domestication and the Development of a Social Bond* (2010). Anyone seeking a solid review of other works in this field dating back to the 1960s should consult chapter 1 of Morey's work. Because Morey's training is in archaeology and anthropology, his book is quite different than ours in both structure and content, although there are thematic congruencies. Morey's book represents a significant advance on Olsen's *Origins of the Domestic Dog: The Fossil Record* (1985), which was at the time of its publication the gold standard for study on the evolution of domestic dogs.

One important book that is the exception to the pattern we described above is very relevant to our theme: *Wolf and Man: Evolution in Parallel*, edited by Roberta Hall and Henry Sharp, two insightful anthropologists who examined similarities between human hunter-gatherers and wolves as an exploration of the "nature-culture opposition in our society" (1978, xi). They state, "Empathy and admiration for our own ancestors and their remarkable histories of survival lead us to a third source of inspiration, admiration for the wolf and its fellow gene pool members, the coyote and the dog." They continue, "Aboriginals admired the wolf and coyote and certainly took no displeasure in their presence, as the remarkably rich traditions of ritual [and] myth . . . among North American native peoples amply illustrate" (xii). Although Hall and Sharp did not consider a cooperative relationship between the species (it was, after all, 1978), they set forth a series of interesting propositions that accorded inspiration for Pierotti when he first began to consider some of the themes we discuss in this book.

Any book that deals with wolves and possible mechanisms for the evolution of dogs must acknowledge Ray and Lorna Coppinger's book, *Dogs: A Startling New Understanding of Canine Origin, Behavior, and Evolution* (2001). This book has attained a degree of popular and media recognition rarely achieved by works with a scientific theme. We are not sure how future scholarship will treat this work; however, we offer a thorough critical evaluation of its basic premise in our text.

Anthropologist Pat Shipman's book *The Animal Connection* (2011) represented a gold mine of early inspiration. Shipman's latest book, *The Invaders: How Humans and Their Dogs Drove Neanderthals to Extinction* (2015), complements our efforts by exploring in detail the archaeological record of Neanderthal people and examining how their disappearance seems closely timed with the arrival of modern humans and their relationships with dogs (wolves). Shipman argues that this pairing made modern humans superior ecological competitors so they were able to supplant the (apparently) "dog-free" Neanderthals.

Belgian archaeologist Mietje Germonpré and her research team have shown fascinating evidence of very early domestication, or at least of associations between humans and wolves that are different than any others ever found. Mietje has been generous in providing manuscripts, published papers, and solid insights to aid our approach.

Another colleague, the Australian anthropologist and ethnographer Deborah Bird Rose, has supplied deep insights into the relationship between Australian Indigenous peoples and dingoes in her books *Dingo Makes Us Human* (2000) and *Wild Dog Dreaming: Love and Extinction* (2011). She has been a continuing inspiration throughout our research.

Dr. Cynthia Annett introduced us to the situation in southern Siberia, where wolflike west Siberian laiki (the plural of *laika* among Russians and Siberians), the dogs preferred by both hunters and herders as companions, roam the streets. Her expertise regarding canid behavior provided us with inspiration and guidance, and her editorial skills refined our prose.

Vladimir Beregovoy, an expert in Russian ancient dogs and laiki, supplied illustration and useful insights into our work. His book *Hunting Laika Breeds of Russia* (2001) is a must for anyone studying this field and was a great help in writing chapter 5.

Our Blackfoot and Anishinaabe colleague Nimachia Hernandez contributed numerous examples relating to our theme from her peoples and helped us determine what was or was not based on solid information in examining Indigenous accounts.

Two colleagues, Dr. Noreen Overeem and her partner Tracy McCarty, have participated with Pierotti in conducting outreach programs using well-socialized wolves. Without the friendship and support of Dr. Overeem and Ms. McCarty, we would not have gotten to know the wolf called Peter, who became one of our primary inspirations in writing sections of this book, especially chapter 11.

We would like to acknowledge the inspiration of two colleagues who are sadly no longer with us. Gordon Smith, who bred captive wolves to produce an apparent contradiction—domesticated wolves—provided insight into human relationships with *Canis lupus* in his unacknowledged classic, *Slave to a Pack of Wolves* (1978). Gordon Haber, who spent more than forty years studying wolves in Alaska and campaigned for their protection against efforts to exterminate them, has shown that humans can continue to live closely with wild wolves, a principle also illustrated in Nicolas Jans's extraordinary account *A Wolf Called Romeo* (2015), the story of a wolf who chose to associate with humans and domestic dogs in Juneau for several years (see the conclusion to this book).

Pierotti would like to express gratitude to the wolves with whom he has shared his life: Liege, who was patient and tolerant with him when he was young and ignorant; Conan, who revealed the workings of interspecies cooperation; Nimma, the social secretary, who showed him how to be independent yet remain an important part of a group; Taba, the wolf scientist, who taught him that wolves could establish social relationships with members of several different species; and Peter, the diplomat and the greatest gentleman he has ever met. He also thanks his current group: Naki Sari, the survivor, who taught him to find vole colonies he didn't know were present and why raccoons can never be trusted; Tosa Kwiisu, the gentle but fierce spirit, who asked to join the pack and learned to rebuild trust; and Tuka Nami, who has shown that being the lowest-ranking dog in a social group does not mean that you cannot have a vibrant and powerful spirit.

Brandy Fogg would like to express her own gratitude to her family: Randall and Jenette, thank you for all the love and support; Derrian, I love you little brother. She also thanks Tyson, aka T (boxer), who taught her that true strength is steady and calm; Tanner (shih tzu), the little hairy dog with a stomach made of steel and her oldest friend, who kept her feet warm from the age of five to the day she left for college; Zar (old English bulldog), who taught her there is always room in the heart for more dogs after a loss; Shyle (shih tzu), the little diva who taught her that alpha means female; and Canis Leonidus, aka Leo (Boston terrier), the fierce protector who taught her patience and offered unwavering companionship as they started a life on their own. This research began thanks to the college students and administrators who allowed her to pick up their dogs for "day care" so that she could conduct the social experiment that started this journey toward understanding canine behavior and inspired further research into the origins of the human and canine relationship.

Finally, we would like to thank all of the Canadian First Nation peoples and U.S. tribes who have stayed true to their cultural traditions and stood strong with their brothers the wolves, allowing safe havens for them on reservations and resisting efforts to exterminate them, even to the extent of lawsuits and exerting social and economic pressure on both state and federal governments.

THE FIRST DOMESTICATION

The Beginnings

Dogs are Wolves who came to live with us.

Tim Flannery, *The Eternal Frontier*

Sometime in the last 100,000 years, a young female wolf found herself alone. She had been exiled from her pack by her mother because she became pregnant and was not the alpha female. If she had stayed with her pack, the older, more "dominant" female would have harassed her continually, finally pushing her aside and killing her puppies at birth; families can care for only one litter at a time. Wolves are the most monogamous of large mammals, and the young female needed to find a partner, or a new family, to help provide for her and the coming puppies. Her old family controlled the area, which discouraged or even prevented young male wolves other than her relatives from finding her and starting a new family group. Like all wolves, the young female was a highly social individual with a need for others, not just as protectors but to share her life.

She recalled a place where there were openings, suitable for a den, in the rock face above a small river valley. Her family had sometimes rested there on hot summer days. Arriving at the valley, she found that a group of humans had built a camp around the openings. Humans, who had arrived only recently, were also a social species, living in groups that seemed not unlike wolf families. They hunted maral (red deer) and bison; she had even seen them scavenging the kills of her family when they first arrived. Her desire for companionship was so strong that she approached the humans but stopped a safe distance away, sitting on her hindquarters, watching. One of them noticed her and gestured to the others. The humans made sounds, but they did not seem frightened or angry, so she moved to one small opening and slipped inside.

Over the next several passings of the sun across the sky, she explored, catching rabbits, grouse, and ground squirrels to feed herself. The humans ignored her for the most part, although their young ones often sat on a nearby rock, watching closely and even following her when she hunted. But they did not interfere or participate.

After another cycle of the moon, her belly was larger, making it more difficult to hunt. She became accustomed to the practices of the humans. They used more types of food than did wolves. Smaller, less hairy individuals that she assumed were females gathered plants and dug up roots; others snared rabbits and grouse. Larger, more hairy ones hunted large prey, usually from ambush; running only on their hind limbs, they were too slow to chase down hoofed animals, but they made a few kills, which fed the entire group.

One day male humans brought back the carcass of a maral hind. The young female wolf watched the females cut apart the carcass, using tools of stone. One younger, slender female pointed to where the female wolf sat in the entrance to her den and made sounds. Another, heavier female severed the neck and head from the carcass and handed them to the younger female, who walked toward the female wolf, whose entire body tensed. The human stopped about twenty paces from the den, laid the piece of the hind on the ground, and made a gesture like she was pulling toward the wolf. Then she returned to the butchering site to work on removing the legs.

The human's action was outside the wolf's experience. She was curious but cautious. Eventually, her hunger overcame her caution; she dragged the head and neck to her den and filled her belly for the first time in weeks. This was almost as good as having a male partner to provide for her.

After that, young human females regularly brought her food, sometimes a rabbit or a fish, enough so that she could rest and prepare for her birthing. They continued to feed her while she nursed her puppies, three females and one male. When the puppies grew teeth and could see, hear, and walk easily, they would leave the den. Young humans and even full-grown females would come to play with the puppies and feed them small bits of meat. It was as if they were all members of the same group.

As the puppies grew, they and their mother became comfortable enough with the humans to wander among them. Occasionally there would be conflict when a puppy tried to steal food meant for a human, but these squabbles were no worse than those that might arise within a wolf group. The humans quickly learned rudiments of wolf behavior from watching the mother wolf discipline her young ones and observing squabbles among the puppies themselves.

In the autumn, when the female began teaching her puppies how to hunt, young humans often joined them. Adult male humans watched this; then some began to participate in the lessons. The collaboration proved mutually beneficial: with wolves as hunting partners, humans could now run down swift prey, and humans reduced the risk of injury to the wolves by helping to finish the kill.

As the days grew shorter and the weather cooler, it was time for the humans to move to their winter camps in the lowlands. The mother wolf and her pups moved with them; they were now one group, living and hunting together. As the years passed, this combined two-legged and four-legged family became the most successful hunters in their lands. When the wolf pups grew old enough to leave their birth packs, they were free to do so—however, they did not fear the humans and in turn the humans did not fear them. Some young wolves joined other human groups, while others formed wolf families. They all shared the land and the herds with their two-legged and four-legged relatives.

This scenario is based upon thirty years of research experience watching wolves and their interactions with humans and with one another combined with careful studies of the traditions and attitudes toward nature of Indigenous peoples in North America (Pierotti 2011a, 2011b; Fogg, Howe, and Pierotti 2015), Australia, and central parts of Asia. Events similar to those described above happened in each of these places, probably occurring many times in some places over the last 40,000 years. Every time modern humans expanded their range into new habitats they found wolves already there (except in Australia; see chapter 6), willing to live alongside them. Many wolves served as teachers and companions (Schlesier 1987; Marshall 1995; Fogg, Howe, and Pierotti 2015).

Wolf packs often have "spare" or "surplus" individuals, usually young adults who are ready to leave the pack but have difficulty finding a mate. There is a fascinating discussion of such individuals and the complexity of pack dynamics in Safina 2015. Such individuals may associate with humans, often establishing friendly, almost mutualistic, relationships. A fictional example of such a wolf is Two Socks from the book and Oscar-winning film *Dances with Wolves*. Recent real-life examples of such individuals are Romeo, a black male wolf in Juneau (Jans 2015; see the conclusion to our book), and the Yellowstone wolves 755 and 820 described by Safina 2015. The story of 820 is particularly poignant. The dominant puppy in her litter, she became a vibrant and charismatic young adult. After the Obama administration removed protected status from wolves, her mother and uncle were killed. Subsequently, her father was supplanted by two younger males who paired with her lower-ranking sisters, who then ganged up to drive her off. She could not pair with her father, so she left the park after her father found a new mate, and like her mother and uncle she was killed by fearful humans.

In cases like the one we describe above, when another female other than the top, or "alpha," female becomes pregnant, this creates stress within a pack or family group (Haber and Holleman 2013, 10–11). When the puppies are born, the lead female will enter the subordinate female's den, groom the puppies and, with apparent regret, kill them. "It is assumed that the killing of pups born to subordinate females by α females may be a common occurrence in both captive and free-ranging wolf packs" (McLeod 1990, 402; see also Safina 2015). Such behavior has also been observed in wolflike breeds of dog; the entire group became agitated and upset over the action, but all members appeared to accept its inevitability (Marshall Thomas 1993).

Infanticide by alpha females, which has been well documented in captive situations, puts exceptional pressure on pregnant subordinate females (Safina 2015). The resulting scenario we describe does not involve humans "steal[ing] wolf pups from their mother's den" or wolves acting as "scavengers on human waste dumps" (Coppinger and Coppinger 2001) or any other of the various models that have been suggested for the origin of one of the closest bonds between different species that has ever existed. This is almost certainly a coevolutionary relationship in which each species influenced the other (E. Russell 2011) to the point that eventually these two species evolved into contemporary humans and our closest nonhuman companions, the domesticated wolves that we now call dogs (see chapters 1 and 9). From our perspective, this coevolutionary relationship developed between two species that found one another compatible and was probably initiated by the wolves, to whom the humans eventually responded in a cooperative fashion. Each species recognized a skill set and emotional capacity in the other that would allow both to maximize their chances of survival and leaving more descendants.

Another difference between our model and those suggested by other scholars is that in assuming a coevolutionary relationship, we do not trivialize either species. According to many Western scientists and their adherents, early modern humans lived in a constant state of terror, huddled around fires at night, trembling if they heard a wolf howl or a lion roar (Francis 2015). This scenario is replicated endlessly on contemporary television, where programs about human evolution provide the latest gloss on the fear-driven model of helpless humans during the Paleolithic. Human hunter-gatherer populations did not, and in the present day do not, live in a state of constant fear. They knew they were (and are) formidable beings that predators attack at their own risk. Elizabeth Marshall Thomas describes one of the clearest contemporary examples of such a situation in Namibia in the 1950s (1994, 2006). The Ju'wasi people, sharing their environment with lion prides, a far more serious threat to humans than wolf packs, dealt with these formidable predators with a mixture of confidence and respect (see also Pierotti 2011a), allowing both species to coexist without fear or hostility. Such affiliations can be described as "treaty making with animal nations" (Simpson 2008). In Africa, with the Ju'wasi, such relationships may go back as far as the origins of modern humans (Marshall Thomas 2006). Early humans dealing with wolves had a much easier time, although the experiences of their African ancestors in dealing with lions may have served them well in establishing a more cooperative relationship with wolves.

Early humans' ability to coexist with predators was an important component of survival skills, and not only in terms of defense—collaboration also made them more effective hunters (Schlesier 1987; Fogg, Howe, and Pierotti 2015). The best examples of such relationships within historical times come from ethnographic studies and the stories and traditions often revealed as a part of such studies. Accounts of interactions between humans and wild canids come from Indigenous peoples of North America, Australia, Siberia, and eastern Asia. Such relationships are based upon mutual respect and the ability to function cooperatively together (Marshall 1995). A European living with Shawnees during the 1770s described the experience as "living with lions" (Gilbert 1989, 20), indicating the powerful nature of these people. Paleolithic humans were not like contemporary Americans, either non-native or native, who are comfortable with smaller, domesticated versions of *Canis lupus,* contemporary dogs. Living off the land, early modern humans preferred canid companions who were equally tough and resilient, considering these nondomestic forms of *Canis lupus* as social partners, credited with "teaching" people to hunt as humans encountered new ecological conditions (Marshall 1995; Fogg, Howe, and Pierotti 2015).

In contrast to Western "scientific" approaches, stories of Indigenous peoples from around the world tell of almost universally positive interactions between wolves and humans (see chapters 5–8). Some stories tell about times when humans, new to an area, survived at first by scavenging kills from wolves (e.g., Schlesier 1987; Marshall 1995; Fogg, Howe, and Pierotti 2015). Most interesting is that in these stories humans are not dominant but take the role of younger members of a society receiving instruction from their elders. In keeping with these traditions, Indigenous peoples repaid their debt over the millennia, leaving parts of kills for the wolves, ravens, and magpies who acted as guides and mentors; in some cases, they even treated wild canids as creator figures (Buller 1983; Schlesier 1987; Pierotti 2011a; Fogg, Howe, and Pierotti 2015).

As far as we can determine, no scholars investigating the origins of domestication have looked to stories or accounts from Indigenous peoples as sources of knowledge concerning human relationships with wolves or of the origins of "dogs." Our goal is to closely examine such stories and knowledge, because Indigenous peoples were the only ones to have actively participated in this dynamic (except Gordon Smith; see Smith 1978). We combine this information with scientific findings to support a model originally suggested by Lopez 1978 and developed by Schleidt 1998 and Schleidt and Shalter 2003.

To understand the way in which we discuss the relationship between humans and wolves, readers should consider the association to be fluid and dynamic—it has changed its nature at different times and in different social and ecological contexts. The relationship began as coevolutionary, with the species cooperating at times but also capable of functioning independently. This state of affairs dominated early stages of the relationship between the two species and may have persisted for 20,000 years or longer. In subsequent chapters we address how in some parts of the world, such as North America and Siberia, this coevolutionary relationship continued until the last few hundred years.

In other parts of the world, for example, southern Asia, humans began to shape wolves into clearly domestic forms: animals phenotypically distinct from wolves, especially in body size (Crockford 2006; Morey 2010). This latter process involves various aspects of the wolf gene pool being essentially divided, with many individuals staying true wolves while others changed in form, becoming what we now describe as "dogs" without losing their genetic links to their wolf ancestry or their ability to interbreed with wolves. The first stages of this separation seem to have begun around 14,000–15,000 years before the present (YBP) (Morey 2010), although no distinct breeds, as these are recognized today, seem to have appeared until the last few thousand years, and the vast majority of breeds current today are at most a few hundred years old. The dingo/New Guinea singing dog phenotype may be the oldest recognizable form that was shaped by human-driven processes. Despite changes in basic phenotype, these animals seem to easily revert to a wild, somewhat "wolflike" way of life, within which they may or may not choose to associate with humans, and they remain fully interfertile with wolves.

As a result of this fluid state of affairs, we consider the relationship between humans and wolves to involve a process during which wolves retained a phenotype not easily distinguished at the physical level from that of wolves that did not interact with humans as well as a somewhat different, more recent process that clearly shows the result of human-driven phenotypic changes: that is, domestication. As Ritvo (2010, 208) has pointed out, "wild" and "domesticated" exist as concepts along a continuum, and the boundary between them is often blurred—and, at least in the case of wolves, it was never clear to begin with.

During the early stages of this coevolutionary relationship, the changes in both species were primarily at the behavioral level. Behavioral traits change very quickly relative to other categories because they are less heritable and more plastic than other traits, especially anatomical features, which show the highest heritability of any class of traits (Roff 1992). Behavioral traits do not fossilize; they can only be inferred, primarily by examining analogous situations in contemporary forms (see below).

Thus, during the early stages of the relationship between humans and wolves, both species probably showed profound behavioral changes without manifesting obvious physical alterations.

All domestication events proceed in this manner: early stages, in which few physical changes are observable, followed after some time by fairly intensive deliberate selection, in which humans prefer specific phenotypes over others. Only during these latter stages are significant physical changes observed. An example can be seen in the fox "domestication" in Siberian fur farms (Belyaev 1979, Belyaev and Trut 1982). Initially, behavioral variation (in personalities) was observed. Such variation undoubtedly had existed for millennia without showing associated physical phenotypic changes. It was only after intense, human-directed selection on behavioral variation had taken place that obvious physical changes began to manifest. *Canis lupus* was clearly the first species to positively interact with humans at a behavioral level and the first to be domesticated by humans under any possible definition of domestication, hence the title we have chosen for this book.

A NOTE ON METHODOLOGY

To effectively examine the evolutionary dynamics of this coevolutionary relationship, our research is a synthesis of data and materials from evolutionary biology, ecology, history, and ethnobiology, which is itself an amalgam of ecology and anthropology (Pierotti 2011b). One master of examining *consilience* between the sciences and history, specifically evolutionary biology and the history of science, was Harvard's eminent paleontologist and natural historian Stephen J. Gould (2003; see also Prindle 2009). In his final book, *The Hedgehog, the Fox, and the Magister's Pox*, Gould laid out a critique of the use of reductionist approaches to address all problems in the sciences. More important, he presented a methodology for examining consilience between the sciences and the humanities, a project he regarded as one of the major enterprises of human thought in the twenty-first century.

We follow Gould in his discussion of how to create synergy between science and the humanities, framing our arguments in terms of history and ethnography mixed with ecological and evolutionary knowledge. Gould argued for the importance of *contingency*, defined as "the growing importance of unique historical 'accidents' that cannot, in principle, be predicted, but remain fully accessible to factual explanation after their occurrence." Such phenomena are significant: "The role of contingency as a component of explanation increases in the same sciences of complexity (e.g., ecology, evolution) that become more and more inaccessible to reductionism" (2003, 202) because of the existence and importance of emergent properties in understanding how these complex systems function. Gould argued that historical uniqueness (contingency) has always been a "bugbear for classically trained scientists" (224), because "unique historical events in highly complex systems happen for 'accidental' reasons, and cannot be explained by classical reductionism."

> If adequate scientific understanding includes the necessary explanation of large numbers of contingent events, then reductionism cannot provide the only light and way. The general principle of ecological pyramids will help me to understand why all [terrestrial] ecosystems hold more biomass in prey than predators, but when I want to know why a dinosaur named *Tyrannosaurus* played the role of top carnivore 65 million years ago in Montana, . . . [or] why marsupial *Thylacines* served [this role] on the island continent of Australia . . . I am asking particular questions about history: real and explainable facts to be sure, but only resolvable by the narrative methods of historical analysis and not by the reductionistic techniques of classical science. (225)

We rely upon Indigenous stories for numerous examples. Although these present historical perspectives, such stories require close reading and interpretation. We combine Gould's approach to historical contingency with analytical techniques used to examine the scientific bases of Indigenous ideas devised by eminent ethnobiologist Eugene Anderson (1996, 103–4):

1. Look for practical information contained within story content.
2. Anything that does not look obviously practical and empirical should be analyzed to see if it is an ordinary, accurate observation described in a culturally unique way.
3. Identify what are actually empirical observations, confirmed by experience but explained by recourse to imaginary constructs.

4. Consider if apparent "errors" can be explained as logical deduction from known principles.
5. Recognize how some counterfactual knowledge creeps into belief systems as the result of teaching devices such as stories because use of myths and fables is effective at teaching morals (and ecological principles) to the young.

Indigenous knowledge represents solid sources of historical information while also displaying firm understanding of ecological relationships within a narrative framework (Pierotti 2011a, 2011b).

Western scholars invariably ignore Indigenous stories, often considering them to be little more than fairy tales or "myths" (Pierotti 2011a; Fogg, Howe, and Pierotti 2015). We argue that these "stories" represent historical accounts of long-standing traditions of Indigenous peoples, particularly in North America, where human cultures evolved in association with large, nondomestic canids. Such stories describe relationships that began long ago and persisted until recent times, although Native American stories do not deal with the exact time when "historical" events occurred. Many such events happened so long ago that they exist "on the other side of memory" (Marshall 1995, 207; see also Pierotti and Wildcat 2000).

Wolves are widely recognized as the first species to be domesticated (Morey 2010; Shipman 2015). Almost all human cultures have *Canis lupus* (mostly dogs) as companion animals, even if they have no other domesticated species (Hemmer 1990; Morey 1994; Shipman 2011, 2015). There is considerable debate about exactly when this domestication took place. Some evidence suggests that it may have happened as far back as 100,000 YBP (Vilà et al. 1997; Derr 2011; Shipman 2011, 2015). In any case, it is clear that for a long time, including up to about 15,000 YBP, the "dogs" that lived with humans were mostly indistinguishable from nondomestic wolves in terms of either their skeletal anatomy or their genes (Morey 1994, 2010; Derr 2011; Shipman 2011).

There is also debate about how many times this domestication of wolves actually might have taken place and in how many different parts of the world (chapter 1). The basic argument seems to be that different lineages of dog arose from wolves at least five times: in Europe, the Middle East, northern Asia, China, and southern Asia (Morey 1994; Crockford 2006; Thalmann et al. 2013). We suggest that such domestication events also occurred in North America, giving rise to the larger "dogs" used as beasts of burden and as hunting companions by many tribes of the northern Plains and Intermountain West. These North American wolflike lineages may now be extinct and may not have contributed much to contemporary dogs, although Eskimo dogs, Native American dogs, and Alaskan malamutes may represent some aspects of this history.

THE WESTERN (REDUCTIONIST) PERSPECTIVE

Apparent in the Western tradition is a significant amount of contradictory speculation concerning the "origins of dogs"—or, as we prefer to consider it, the initiation of relationships between humans and wolves, because the former is typically framed entirely within a Western ideological and social perspective. A classic example of such thinking is found in a recent book exploring this topic: "What makes this amazing evolutionary story even more remarkable is that for many thousands of years, all wolf-human interactions were overtly hostile. We competed fiercely for the same prey and probably killed each other at every opportunity" (Francis 2015, 25). Similar, although less extreme, arguments can be found in Coppinger and Coppinger 2001, and they reflect a bias toward contemporary European and Euro-American ways of understanding the world (Pierotti 2011a; Medin and Bang 2014).

There is also a bias toward assuming a single origin of domestic dogs. Some geneticists argue that all domestic dogs originated from east Asian stock (Savolainen et al. 2002). A problem with this argument is that fossils of what are now considered to be "domestic dogs" have been found that date considerably further back in time than any remains found in east Asia (Morey 1994, 2010; Germonpré et al. 2009; Ovodov et al. 2011; Shipman 2011, 2015; Thalmann et al. 2013). In response to critiques, Savolainen's group has more recently argued for multiple origins within Asia (Niskanen et al. 2013), but this still fails to address evidence from European sites (Thalmann et al. 2013). As far as we can determine, the bias for seeking single origins of dog domestication lies in the idea that if multiple origins are acknowledged, this is a death knell for the idea that dogs are a separate species or even a subspecies. In modern systematics it is difficult to argue that a species is polyphyletic—that is, has multiple origins—because each lineage may have pursued a different evolutionary pathway (Pierotti 2012b). This also creates problems for conservation biology. Some anti-conservation elements, all Euro-American, have argued that if wolves are the same species as dogs, then they should not be protected because dogs are very common and not endangered at all (O'Brien and Mayr 1991; Coppinger, Spector, and Miller 2010; Pierotti 2011a).

We have found during our research, both in the field and in reviewing the literature, that most of the people who write about or study dogs know little or nothing about wolves, and the opposite scenario seems equally true as well. Most scholars who study "dog evolution" concentrate on genetics or archaeology. The few investigators who study dog behavior seem to know even less about wolves than do the archaeologists who examine this topic, for example, Coppinger and Coppinger (2001) and Miklósi (2007). Investigators who work with canid genetics rely upon evidence from DNA molecules rather than observation and seem to believe that genotypes trump phenotypes in understanding evolutionary history. Behavioral investigators seem to assume two major factors: (1) initial relationships between wolves and humans were competitive rather than cooperative (Coppinger and Coppinger 2001; Francis 2015), and (2) humans had to have taken the initiative in establishing the relationship. This combination of ideas seems to generate further European-influenced bias because, like the majority of people of European ancestry, such researchers cannot imagine that other species may have been superior to humans in skill sets, or perhaps even dominant in early phases of the relationship. This expresses itself in terms of both superiority and fear (see the quote from Francis 2015 above). Even one of the most insightful scholars working in this area (Shipman 2011, 2015) regularly uses such terms as *aggressive, dangerous,* and *ferocious* to describe wolf behavior. A similar attitude can be seen in the discussion of wolf behavior by the Coppingers (2001), in which Ray Coppinger describes how he was "attacked" by a wolf at Wolf Park (see the discussion of this incident in chapter 1). Such ideas seem rooted in the idea, set forth in Genesis, that humans have "dominion over the world." Moreover, it is assumed that humans never have positive relationships with adult wolves. As Morey (1994, 2010) has pointed out, most Euro-Americans have difficulty separating the strong symbiotic relationship that modern humans have with domestic animals from the processes that led to their origins in the first place.

In contrast, terms Indigenous peoples use to describe wolves are *brother, grandfather, relative, companion, teacher,* and even *"creator"* (Ramsey 1977; Buller 1983; Schlesier 1987; Marshall 1995; Pierotti 2011a, 2011b; Fogg, Howe, and Pierotti 2015), which suggests strongly that Francis's idea that "all wolf-human interactions were overtly hostile" is at best simplistic and at worst an example of what we might refer to, to coin a phrase, as *Euro-bias:* the idea that only experiences of Europeans and Euro-Americans count in assessing ecological and social relationships between species (see chapter 7). Indigenous cultures all had dogs of varying degrees of domesticity prior to the arrival of domestic dogs accompanying Europeans as they invaded America. The obvious difference in attitudes between Indigenous sources and European and Euro-American scholars suggests that it might be worthwhile to examine how these non-Western cultural traditions describe their experiences in dealing with wolves and their understanding of how the transition from wolf to dog took place (e.g., McIntyre 1995; Coleman 2004).

We are not attempting to explain the origins of the hundreds of "breeds" of dog that exist today (Morris 2001; Spady and Ostrander 2008; Hunn 2013). Most breeds are recent creations of humans, having been developed only in the last two centuries, and many breeds are of even more recent origin, such as the recent trend toward "designer" dogs, new breeds created by mixing established breeds, such as "peekapoos," crosses between miniature poodles and Pekingese. Our intention is to discuss the initial process by which humans and wolves established a coevolutionary relationship that eventually resulted in the creation of the canids we call "domestic dogs." A trap that some geneticists have fallen into is the attempt to determine the genetic relationships among modern breeds (vonHoldt et al. 2010). There has been so much interbreeding among various lines of domesticated or semi-domesticated wolves (dogs and wolf-like dogs) that although it may be possible to establish recent breeding histories, little other than general categories is revealed through these studies (Coppinger, Spector, and Miller 2010; Pierotti 2014). In addition, some crucial breeds, such as the large dogs maintained by Indigenous Americans, no longer exist or have been absorbed into the overall "dog" gene pool (Fogg, Howe, and Pierotti 2015).

Our line of thinking provides a possible explanation of why many Russian "wolves" and "dogs" are more difficult to distinguish using either

DNA (Vilà et al. 1997) or anatomy (Germonpré et al. 2009). In central Asia, wolves and "dogs" have remained similar and continue to interbreed. Many dogs that exist in Siberia today are difficult to distinguish from wolves, and indeed would be considered to be wolves by contemporary Americans, including many presumed experts in this field (see chapters 5 and 10).

This privileging of Western concepts and the inability to imagine relationships between wolves and humans in other than contemporary Western terms (Euro-bias) leads to a number of difficulties. Both behaviorists and geneticists end up arguing about the "origin of domestication" as if it were a single event that happened only once (e.g., Coppinger and Coppinger 2001; Savolainen et al. 2002; vonHoldt et al. 2010). Both groups fail to recognize that there are at least two crucial questions involved in understanding domestication and the relationships between *Homo sapiens* and *Canis lupus*. First is the issue of when this event initially occurred, which may never be clearly established, although some archaeological evidence suggests that this happened at least 30,000–40,000 YBP (Germonpré et al. 2009; Ovodov et al. 2011; Pierotti 2014; Shipman 2015; but see Morey 2010, 2014 for a differing perspective). Second, none of these scholars address the issue of relatively recent doglike wolves (or wolflike dogs) that existed until at least the nineteenth century in many areas, and may still exist in some parts of the world today, for example, central and eastern Siberia and North America (Beregovoy and Porter 2001; Beregovoy 2012; Pierotti 2014; Fogg, Howe, and Pierotti 2015). This second issue is especially important because if humans continued to domesticate or at least have close social relationships with various forms of *Canis lupus* until recent times, this makes the issue of when and where such events first happened of little relevance. Skeletal remains identified as dogs, rather than as wolves, have been found in Europe that date back to 32,000 YBP (Germonpré et al. 2009; Shipman 2011, 2015; Thalmann et al. 2013). In central Asia similar remains date back to at least 33,000 YBP (Ovodov et al. 2011). Domestication of wolves appears to have taken place in various locations at various times, including Europe, the Middle East (Levant), central Asia, eastern Asia, and probably India and North America (Morey 1994, 2010; Savolainen et al. 2002; Verginelli et al. 2005; Ovodov et al. 2011; Skoglund, Götherström, and Jakobsson 2011; Thalmann et al. 2013).

A major complicating factor is that within the last few hundred years humans have bred earlier types of dogs into contemporary lines and vice versa (G. K. Smith 1978; vonHoldt et al. 2010). Combined with periodic introgression from wild wolf populations, especially in ancient, primitive, or aboriginal breeds (G. K. Smith 1978; Beregovoy 2012), this has resulted in many different categories of dog that cannot be linked to a single origin (Coppinger et al. 2010; Pierotti 2014). We examine how this domestication could have begun and then expanded over time, with special attention to central Asia, Japan, Australia, and North America, where Indigenous peoples have very different relationships with wolves and wolflike dogs (chapters 5–7). Where the stories of Indigenous peoples deal with this relationship, we explore how this process continued until recent times in an effort to understand the origins of the relationship that humans have today with domestic dogs.

There are many reasons that gray wolves (*Canis lupus*) were the first species to be domesticated by humans. Wolves share with humans a similar family structure and accept joining human groups more easily than species that do not live in large extended social groups, such as coyotes (*Canis latrans*) (Lopez 1978; Schleidt 1998; Haber and Holleman 2013; see chapter 3 for extended argument). Modern humans probably learned early on in their evolutionary history of the benefits that came from sharing their lives with wolves (Schleidt 1998, Schleidt and Shalter 2003; Shipman 2014, 2015).

Part of this debate centers around the difficulty in distinguishing early dogs from wolves, because they are virtually indistinguishable at the skeletal level (Morey 1986, 1994, 2010; Franklin 2009; Shipman 2011). Scientists often use morphological traits of the skull to distinguish wolves from dogs, but this approach is fraught with problems. Many modern-day dog breeds are very wolflike in their skeletal structure and behavior, and these breeds are likely to be similar to the earliest dogs (Morey 1986, 2010). For most of their history the wolf phenotype may have been the basic form for a domestic animal that primarily functioned as a hunting companion for early humans.

Despite debate about which species is the ancestor of domestic dogs, the conclusion of modern scientists seems unequivocal: "Although there are clear differences between the species, the dog matches the wolf better than either the coyote or the jackal" (Hemmer 1990, 38). It is clear that genetically wolves and dogs are very similar (Vilà et al. 1997; vonHoldt et al. 2011; Thalmann et al. 2013). The modern family Canidae evolved in the Miocene of North America about 6 million years ago (R. Nowak 1979; Wang and Tedford 2008). The line that led to *Canis lupus* migrated to Asia through Beringia during the Pliocene, evolved into *Canis lupus* in Asia, and then returned to North America again through Beringia during the Pleistocene (Wang and Tedford 2008). This pattern makes biogeographic sense because the gray wolf is the only large canid that lived in all of the places where dogs are presumed to have originated. In contrast, the coyote (*Canis latrans*) and its relatives the red wolf (*C. rufus*), the tweed wolf (*C. lycaon*), and the dire wolf (*C. dirus*) evolved in North America and never left, which means they cannot be ancestors of modern domestic dogs (P. J. Wilson et al. 2000, 2001).

Another theme we discuss is the marked contrast between the attitude of Euro-Americans and Indigenous peoples in North America toward wolves and coyotes. This contrast is made clear in the following quotation from a professional wolfer about his time during the winter of 1861–62 in western Kansas poisoning wolves: "In the days of the Buffalo, wolfing was a recognized industry. Small parties . . . used to go to the Buffalo range, establish a camp, and spend the winter there, killing buffalo and poisoning the carcasses with strychnine. The wolves that fed on these carcasses died about them, and their pelts were taken to camp, to be stretched and dried. . . . The Indians were bitterly opposed to the operations of these wolf hunters, who killed great numbers of buffalo for wolf baits, as well as elk, antelope, deer, and other small animals" (Grinnell 1972). From the European and Euro-American perspective, wolves were a species to be exterminated and no method was too cruel or inhumane. This is a far cry from the way Indigenous groups who opposed these operations saw the wolf and its role in their world.

According to many Plains peoples, wolves live lives that parallel those of humans, and thus represent another people that must be respected. The Blackfoot, Lakota, and Cheyenne (Tsistsista) claim to have learned to hunt from wolves (Schlesier 1987; Marshall 1995, 2001; Pierotti 2011a; Hernandez 2014; Fogg, Howe, and Pierotti 2015). It is important to emphasize that for people who live by hunting, being taught new ways to hunt is tantamount to an act of creation, for this learning caused cultures to be fundamentally reshaped (Pierotti 2011a).

We examine traditional stories from Indigenous peoples around the world to see how they describe relationships between wolves and humans. In some cases, such stories are supported by the accounts of early European observers; an eighteenth-century Canadian explorer reports: "Few Northern Indians chuse to kill the wolf, under a notion that they are something more than common animals. . . . I have frequently seen the Indians go to their dens, take out the young ones and play with them . . . and I have sometimes seen them paint the faces of the young wolves with vermillion or red ochre" (Hearne 1958, 224). Such accounts suggest that Indigenous peoples deliberately formed relationships with wolves. Indigenous peoples may have identified particular young wolves that might make good companions, which would explain the face-marking procedure. The events described by Hearne seem to indicate the establishment of relationships between individual hunters and wolves and suggest that wolves may have continued to be brought into human groups even after the arrival of Europeans.

The majority of changes that took place during the process of domestication were behavioral and would not be seen in morphological remains. Indigenous groups probably initially benefited from their relationship with wolves; however, with the passage of time, the role humans wanted wolves to play in their lives changed and morphological changes became more evident. It seems likely that "even though the behavioral changes driving morphological shifts [in dogs] are not visible, the neoteny of the skull is preserved" (Schwartz 1997, 10). Skulls incorporate important features that allow scientists to determine if a fossil is a wolf or a dog, because mammalogists look primarily at skulls in assessing taxonomic status, typically assuming that the facial shortening that the wolf undergoes in becoming a dog is a key criterion (Morey 1994, 2010; see also Schwartz 1997). This "shortening" involves reduction of muzzle length along with increasing height of the eyes, producing a stop, or "forehead," in most dogs that is not seen in wolves.

Once humans realized that dogs could be changed, certain breeds, or body forms, were developed to perform specific tasks more efficiently, and thus humans selected for specific traits such as broader shoulders for pulling travois or sleds. Individual wolves that made the transition to becoming domesticated animals became more easily controlled by humans. In contrast, wild wolves that avoided domestication remained completely independent.

ORGANIZATION OF THE BOOK

In chapter 1, we discuss our interpretations of the meanings of the concepts of "wolf," "dog," and "wolf-dog" from an evolutionary perspective. This is important because Raymond Coppinger and his colleagues at Hampshire College have muddled these concepts in an essay titled "What, if Anything, Is a Wolf?" Such an approach illustrates a basic weakness in the Coppingers' 2001 book, that is, little understanding of contemporary systematics and evolutionary thinking on the nature of species and the role that hybridization can take in shaping the evolution of closely related forms. Despite major weaknesses, this work has drawn considerable attention, being showcased in popular media, including *National Geographic* and PBS, and it is important to evaluate the Coppingers' argument.

The title of chapter 1 is a gloss on a well-known article in evolutionary biology, "The Spandrels of San Marco" (Gould and Lewontin 1979), which discusses why it is often a mistake to try to find an adaptive explanation for every aspect of an organism's biology. Much literature on dogs shows similar simplistic, often mistaken, assumptions concerning the processes by which dogs evolved from wolves. Most people expect that there is some clear line that exists between the taxonomic categories "wolf" and "dog," a problem that began when Linnaeus classified domestic dogs and gray wolves as distinct species in the eighteenth century. A classic example is the Coppingers' 2001 book, which contains numerous errors of logic in discussing the evolution of dogs and interspecies relationships. We conclude by showing that each human cultural tradition developed with specific images of the canid (or possibly canids) that was able to share their particular way of life. This is why it is so hard to define the term *dog*—different cultural traditions, and even individuals within these traditions, have very different images of what type of canid is best suited for relationships with humans.

Chapter 2 reviews the study of cooperative behavior between species, with emphasis on examples of cooperative hunting found in a wide range of species. Seen in this context, the idea of cooperative hunting between humans and wolves that evolved into our present relationships with dogs does not seem unusual or surprising. We critique the proposal that competition between species is more important than cooperation in structuring ecological communities, discussing how this notion leads to a suite of ideas philosophically separating humans from the rest of the natural world. In many ways Western science is unintentionally complicit in such thinking. We close by discussing complex cooperation, including long-term relationships between members of different species.

Chapter 3 examines what it means to be human, a member of the biological species *Homo sapiens*. We explore the essence of our biological nature and attempt to evaluate the biological and cultural pressures that have shaped humans into the social mammals we are today. Comparing humans to a wide range of primates, we show that no other species, and certainly none of our anthropoid relatives, has a similar social structure, with social groups of varying sizes built around nuclear families. Finally, we explore how these traits, especially the cultural ones, may have been shaped by our shared experience with *Canis lupus*. In this chapter, we explore the implications of an idea from Schleidt and Shalter (2003), supported by Haber and Holleman (2013), concerning the coevolution of humans and canids. We examine the history of Western philosophical attempts to define humans as separate from all other organisms, and show how such ideas are invariably rooted in creationist-type thinking. Humans are indeed unique, but our adaptations emerge from a set of unusual events, and a considerable amount of the history of modern human evolution seems to be influenced by our association with wolves and their dog descendants. We close by showing how modern attitudes toward predators result from religious traditions rather than scientific understanding.

In chapter 4 we review archaeological research and its role in explaining the transformation from wolf to dog, addressing why this topic is controversial: the tendency to identify wolf remains found in archaeological sites as evidence of either interlopers or human killing overshadows the alternate possibility of social bonding between humans and wolves. This probably has prevented appreciation of considerable early evidence of relationships between humans and wolves before the latter became sufficiently phenotypically distinct ("doglike") to be recognized as domestic animals shaped by humans. Some archaeologists do not acknowledge the possibility that humans interacted with and coevolved with wolves for thousands of years without generating significant phenotypic change in either species, and thus early wolves living with or cooperatively hunting with humans probably go unrecognized by scholars looking only at obvious physical changes. From our perspective, the recent controversy among contemporary archaeologists results because different groups are addressing different, albeit linked, questions, especially in North America, where humans retained close social relationships with nondomestic wolves up until at least the nineteenth century.

In chapters 5–8, we examine relationships between wolves, wolflike "dogs," and humans, starting with the history of humans and canids in Asia. The history of Western civilization reveals a long-standing tradition of demonizing or dehumanizing other peoples, especially peoples considered as potential rivals for territory or resources. In contemporary culture, this attitude manifests itself in the entertainment industry's obsession with "werewolves": the hybrid nature of such creatures can be seen as an example of one of the most consistent bugbears of "civilized" nations and societies—dog-men or cynocephali. The location of these fearsome hordes has always been the far reaches of uncharted "wilderness." Such beliefs are based upon practices among tribes in central Asia: the men hunted with wolves or large "wolflike" dogs and at times wore masks or capes of dog skin when fighting. Similar traditions continue to this day in Siberia: men spend much of the year in hunting camps, accompanied by laiki, a wolflike primitive dog breed. DNA studies of Russian wolves and dogs show little clear division between wild and domestic canids. In Japan until the late nineteenth century, humans enjoyed a basically cooperative and benign relationship with local wolves. The identities involved are not clear because traditional Japanese describe "wolves" as benign, whereas they are more cautious about what they call "mountain-dogs" (Walker 2005).

In chapter 6 we discuss the distinctive situation in Australia, where *Homo sapiens* and dingoes coexisted for several thousand years. These two species were the only large placental mammals on a continent dominated by marsupial mammals and large reptiles. Considerable mythology has developed around the relationship between humans and dingoes (Rose 2000, 2011). The dingo group represents a unique branch of canid domestication; they live independently, either wild or semi-wild, but also associate with humans, including hunting and even sleeping with them. Dingoes demonstrate that it is possible for an animal previously domesticated (shaped by humans) to live and reproduce successfully without humans. The relationship between dingoes and the Aboriginals provides a model for investigating the process of domestication in canids, strongly suggesting that domestication is a multistep and potentially reversible process. It also reveals how a canid can exist in a situation where it may or may not choose to live with humans, functioning well in either case.

Chapter 7 examines the historical relationship between Indigenous Americans and wolves illustrated through the stories of Indigenous peoples of North America, especially on the Great Plains and the Intermountain West. Tribal accounts have not been previously employed in scholarly examinations of the origins of "dogs" or studies of domestication. All the Plains tribes examined closely (Cheyenne, Lakota, Blackfoot, Pawnee, Shoshone) have stories characterizing wolves as guides, protectors, or entities that directly taught or showed humans how to hunt, creating reciprocal relationships in which each species provided food for the other or shared food. The Coppingers (2001) argue that the first wolves associated with humans scavenged or hung around camps waiting for scraps, and thus the process of domestication began with wolves being dominated by humans. In contrast, evidence from tribes suggests a coevolutionary reciprocal relationship between *Homo sapiens* and American *Canis lupus* that existed until at least the nineteenth century.

Chapter 8 discusses an intriguing aspect of the relationship of humans with wolves in North America and parts of eastern Siberia—that wolves are considered "creator" figures, suggesting that they played an important role in the way humans conceived of themselves as they adapted to new environmental conditions. Thus, wolves could function as both teacher and creator to peoples who were willing to respect wolves as hunters and pay attention to the examples they set. A related trope, often confused with the creator figure, is the idea of smaller canids such as coyotes and foxes as "trickster" figures. The concept of "creation" in Indigenous non-Western cultures differs from its meaning in Western religious tradition. The former describes the origin of a new cultural tradition in metaphoric form, whereas the latter attempts to represent some unknowable event such as the origin of the universe, the earth, or of life. Following Lévi-Strauss 1967, we discuss why tricksters among many American tribes are scavengers and omnivores, for example, coyotes and ravens, occupying an ecological mediating position between herbivores and carnivores and "in between" in terms of subsistence strategies. Humans are also omnivores and in their early history probably functioned as scavengers.

Chapter 9 explores the history of the process of domestication conceptually and considers the meaning of the terms *domestic, wild,* and *tame* and how the concept of *feral* fits within this framework. We explore differences among traits in generating phenotypes, including the classic Russian studies on fox behavior and morphology. Domestic dogs are not a natural grouping because they involve multiple lineages (polyphyly) that have undergone extensive interbreeding among lines (reticulate evolution), which reduces clarity concerning traits that might be used to identify dogs as a species. In particular, some recognized "dogs" resemble their wild ancestors quite closely, whereas other types of "dog" bear little resemblance, raising questions of the role of neoteny and differential developmental rates in canid evolution. Evidence of polyphyletic origins is also found in other domesticated animals, especially cats, cattle, and pigs. Differences between what it means for an animal to be *tame* versus *domesticated* reveals that these concepts are regularly confused and conflated in the popular literature, even by scientists. This leads to examination of the use of rabies vaccines in different animals and why this debate has produced unscientific, creationist-style thinking, even among supposed evolutionary biologists. For many breeds of larger dog, distinctions between "dog," "wolf," and "wolf-dog" have little meaning.

In chapter 10, we discuss how many types of "dog," including AKC-registered breeds, can be mistaken for wolves by people who have stereotypical ideas of what a wolf is, including some wolf biologists. We deconstruct the concept of "experts" who attempt to make such distinctions, including those who advise state and federal legislators. We also explore relationships between humans and wolves (both domestic and nondomestic) as social companions by evaluating Pierotti's experiences as an expert witness distinguishing between dogs and wolves. This type of confused thinking has led to bad laws based upon emotional responses rather than attempts to govern effectively and write laws that improve the functioning of society.

In chapter 11 we examine our own experiences and those of colleagues who have lived for many years with crosses between wolves and dogs, and even pure wolves, and their perceptions of the bonds and relationships that exist between humans and different types of canids. We have dealt with a couple of dozen wolves, around a hundred crosses between wolves and dogs, and a couple of hundred dogs in various capacities, including as owner, companion, scholar, mentor, trainer, educator, and expert witness. A crucial point is that the social bond between humans and wolves that changed into domestic dogs is the source of both major pleasures and major conflicts between humans and their canid companions. Large domestic dogs have the anatomy of serious predators combined with a confidence in their interactions with humans that can lead to aggression and grave conflict. In contrast, wolves and high-percentage crosses between wolves and dogs tend to be timid, retreating when faced with unfamiliar humans. We discuss the "danger" presented by various breeds, including wolves and wolf-dogs, and challenge a number of points of received thinking, including the notion of the equivalency of "wild" and "dangerous." A major aspect of the "danger" from a canid is associated with size above all else, which is to be expected in dealing with large predatory animals.

Our conclusion follows up on this last point, addressing the great enigma of the first domestication: wolves and dogs are so affectionate and seem willing, if not driven, to create strong and persistent social bonds that it becomes easy for humans to anthropomorphize and idealize these four-leggeds that share our lives so easily. Yet they remain predators, highly evolved carnivores, and they know how to kill. Given the opportunity, sometimes they kill things that humans value, for example, other domestic animals. As long as humans considered themselves to be fellow predators, however, we lived comfortably with wolves. They were our companions, sharing both our hunts and our kills and living with us in a more or less equal sort of reciprocity. We conclude by examining one particular case of a wild wolf that showed repeated friendly relations with humans over a period of several years and discuss the implications of such experiences.

CHAPTER ONE

The Spaniels of San Marcos

WHAT IS A DOG AND WHO CARES?

"ALL DOGS ARE WOLVES, but not all wolves are dogs." This simple concept has profound impacts for understanding the evolutionary history of what we think of as "dogs." Most people expect some clear line to exist between the taxonomic categories "wolf" and "dog." This misunderstanding began when Linnaeus classified domestic dogs as *Canis familiaris* (familiar dog) and gray wolves as *Canis lupus* (wolf-dog) in the eighteenth century (Linnaeus 1792). Separating wild and domestic forms (see chapter 9) was typical in Linnaeus's classification scheme because he did not think in evolutionary terms; as a standard-issue eighteenth-century creationist, he was trying to comprehend the "mind of God." As a result, he considered the wild and domestic forms to have each been specially created, even if they appeared similar enough to be linked within his classification scheme. We do not know which specific breed of dog Linnaeus used as the type specimen for *C. familiaris*, so it is unknown which phenotype the taxon *C. familiaris* was based upon. It is even possible that the breed of dog that Linnaeus based *C. familiaris* on no longer exists (E. Russell 2011). This is not a problem when discussing the Linnaean taxon *C. lupus*, because nondomestic gray wolves are a true species having a consistent phenotype that allows them to be identified on the basis of a type specimen. Domestic "dogs," *C. familiaris,* cannot be so identified. In keeping with modern systematics, the American Society of Mammalogists clarified this misunderstanding by combining domestic dogs with their wild ancestors (see Chris Wozencraft's section on carnivores in D. E. Wilson and Reeder 1993), creating a single taxon *Canis lupus,* including *Canis familiaris,* which is no longer recognized as a species.

Unfortunately, when Wozencraft combined dogs and wolves into a single species, he introduced another source of confusion by listing *familiaris* as a "synonym" for *lupus* in his examination of carnivore systematics (as did Honacki, Kinman, and Koeppl 1982 a decade earlier). This decision was intended to represent the historic taxonomic record (standard practice in systematics), which lists as synonyms all other Latin binomials under which various forms of *Canis lupus* have been classified over the last 250 years since Linnaeus, that is, other historically recognized scientific names for species that we now recognize as being included within *C. lupus.* Other synonyms listed were the names of recognized wolf subspecies. The listing of *C. familiaris* as a synonym, however, led many people, including some prominent biologists, to the erroneous impression that this decision meant that domestic dogs were now considered to be a subspecies of wolf, *Canis lupus familiaris* (see Coppinger and Coppinger 2001; Safina 2015).

The root problem is the idea that there is really no definable taxonomic entity (species or subspecies) that can be identified as "the domestic dog," characterized by a specific set of physical, behavioral, or physiological traits that differentiate it from wolves and other canids. The sole criterion for identifying domestic dogs is that they be some domestic form of *Canis lupus,* including entities as diverse as bull mastiffs, Chihuahuas, Alaskan malamutes, Siberian laiki, basenjis, cocker spaniels, and shar-peis. This does not reflect the current state of biological and evolutionary reality.

A major problem confronting the thinking employed by people who want dogs and wolves to be different species, or at least subspecies, is that large wolflike "dogs" (malamutes, Siberian laiki, etc.) as well as the highly derived, highly divergent modern breeds (dachshunds, miniature poodles, etc.) are all defined as "domestic dogs," leaving no clear idea of exactly what is meant by the term *dog.* For example, a type specimen was described as *Canis pomeranus* by Johann Friedrich Gmelin for the Pomeranian breed in his revision of Linnaeus's *Systema Naturae* in 1788 (Linnaeus 1792). Although this was a valid name, no one has suggested that *Canis lupus pomeranus* be employed as a subspecies name for this important lineage of dogs. Thus, if domestic dogs are a valid species, as argued, *C. familiaris* has no defined characteristics other than "domestic version of *Canis*," in which case the species identity depends entirely on the definition of *domestic,* a term that is nowhere near as simple to understand or define as many might imagine (see chapter 9 for extensive discussion).

One key point that it is important for readers to keep in mind throughout this book is that whether or not an animal is considered to be a dog or a wolf is not determined by the way it acts. Despite statements made by Coppinger and Feinstein (2015) and other "dog experts," there are today, and probably always have been, nondomestic wolves that behaved in a social and companionable manner toward humans. There are also many more fully domesticated dogs that are unfriendly and aggressive toward humans. These dogs may or may not superficially resemble nondomestic wolves, but it is not their "aggressive nature" that makes them "wolflike."

FOLK VERSUS BIOLOGICAL CLASSIFICATION

It has been argued that "dog" can function either as a folk generic taxon or as a "life-form" taxon, depending upon the frame of reference (Hunn 2013). Under this logic, the term *dog* is functionally equivalent to the Linnaean taxonomic category of genus, equivalent to *Homo* in the case of humans, a category that includes obviously different species such as Neanderthals and Peking Man, *H. erectus* and *H. sapiens.* The category of dog, for example, spitz, herding, retriever, sight hound, spaniel, mastiff, terrier, scent hound, would then function as the equivalent of a species in Linnaean terms (categories from vonHoldt et al. 2010). Specific breeds, for example, cocker spaniel or German shepherd, within these categories are then functional equivalents of "subspecies" (varietals, as described by Hunn 2013). Thus, in the minds of most people, the category *dog* is conceptually equivalent to the genus *Canis;* the functional type of dog, for example, retriever, is equivalent to the species *lupus;* and the breed, for example, cocker spaniel, is the equivalent of the subspecies; for example, *nubilis* is the subspecies name in the case of the Great Plains, or buffalo, wolf, *Canis lupus nubilis.* In true systematics, however, members of distinct genera are assumed to not be interfertile. For example, members of the genus *Canis* are not interfertile with the genus *Lycaon* (African wild dog) or the genus *Cuon* (dhole or Asiatic wild dog). In reality, categories and breeds of dog are all interfertile, not only with one another but with wolves, their nondomestic *Canis* ancestors.

The creation of the folk category of dog and its subcategories has resulted in a huge amount of pseudo-scientific confusion. The average person, not conversant with systematics or even with Linnaean taxonomy, simply reverts to folk taxonomy, as characterized by Hunn 2013. Unfortunately, there are real-world consequences to this confusion because legislative bodies, animal control agencies, courts, and the general public feel confident in their ability to tell the difference between a wolf and a modern breed of dog. In reality, they have no idea of how to distinguish between wolves, wolflike dogs, aboriginal dogs, ancient dogs, crosses between wolves and dogs, or even between wolves and well-recognized but less familiar breeds, such as Belgian sheepdogs, a situation that creates legal, social, and emotional chaos.

Adding to the confusion is the way in which Euro-Americans conceive of the relationship between wolves and dogs. By using terms like *wolf-dog*, it seems that most people think of such animals as belonging to a "breed" equivalent to French poodle. In their minds, "wolf" is a category of "dog," rather than the other way around. Following this logic and employing the schematic argued by Hunn (above), this makes wolves equivalent to a breed category, like terrier or hound. The logical flaw should be obvious: wolves are the ancestors of all dogs; therefore, their offspring, when crossed with various breeds of dog, cannot simply be considered as another breed, especially because they are typically crossed with large breeds of dogs that resemble wolves, such as malamutes or German shepherds (Morris 2001).

This situation is made worse by a wide range of "experts" who claim to be able to distinguish wolves and wolf-dogs from "dogs" and are often employed by animal control entities to make such identifications (see chapter 10). Erich Klinghammer, who operated a training program for such "experts" out of his privately run Wolf Park in Indiana, is a good example. Klinghammer charged from $350 for a weekend course to $500 for a five-day course to train people to become "experts" in wolf (and by implication wolf-dog) behavior. We seriously question whether people can be validly certified in such complex fields as animal behavior and the nuances of wolf and dog breed-specific behaviors within such short periods of time. Yet many people Klinghammer trained list "volunteer scientist" at Wolf Park on their résumés, including many new age, "holistic" laypeople (a brief search on Google reveals dozens of such individuals).

Although Klinghammer is well known in the field of wolf behavior, his methods were controversial, largely because he released bison into enclosures with captive wolves to demonstrate predatory behavior to audiences (http://wolfpark.org/animals/info/bison/). This practice is not done at any other facility, and did not meet the basic standards of animal behavior research protocols (Animal Behaviour Society 2012).

One of the clearest examples of Klinghammer's confusing influence is an event described in Coppinger and Coppinger 2001. This incident, related in the first person by Ray Coppinger, illustrates problems we describe throughout this book, so we quote it in its entirety, after which we deconstruct it:

> In fact a wolf that is unafraid of humans is more dangerous
> to humans than is a wild wolf. A wild wolf will flee when
> approached, but a tamed one is not afraid to move in and bite.
> Twenty years ago [Klinghammer] took me into the main cage
> with the "socialized" wolf pack [at Wolf Park]. All wolves had
> been born in captivity for several generations, hand raised
> as puppies, and "tamed." All were part of Wolf Park's
> demonstrations and were handled daily. Why was I so
> reluctant? . . .
>
>   "Just treat them like dogs," Erich told me authoritatively.
> Which I did, by thumping Cassi on her side, and saying
> something like "good wolf." That was when she became all
> teeth. Not a nip, but a full war—a test of my ability to stay on
> my feet and respond to Erich's excited command, "Get out! Get
> out! They'll kill you." . . . I had a blurred vision of a collection
> of wolves gathering, and a wolf tugging on my pants as Cassi
> focused on my left arm.
>
>   "Why did you hit her?" Erich said. . . . "It wasn't hitting! I
> was patting her! You said treat them like dogs and I pat dogs
> and if I do some social misconduct with a dog, I don't get my

head bit off, and why is it that all you people who socialize wolves have these nasty scars?" I said in a single breath while applying a tourniquet to the mangled arm of my goose down jacket. Never again did I think that tame wolves could be treated like dogs. (44)

Compare this account to Donald McCaig's imagined encounter with a famous border collie in the Scottish Highlands: "I understood too well that he, dog, was in his element here and that I was the interloper. 'You will not pat my head.' After his initial confident perusal, Sirrah's [the dog] request betrayed a milder aspect and I hastened my assurances. He was reluctant to take me at my word. 'You Americans are the worst. You cannot encounter a dog, going about its lawful business, without stopping to paw at it. If you fondle me, sir, I will not be responsible for the consequences' " (1991, 112). McCaig, a layperson, understands what Ray Coppinger, a scientist and university faculty member, does not—you do not pet (let alone thump) an unfamiliar canid without care. Sirrah (a domestic dog) sends the same message as Cassi (a semi-domesticated wolf) did: "If you fondle me, I will not be responsible for the consequences." To McCaig, the independence and integrity of Sirrah are major aspects of his identity. To Coppinger, his own lack of control is frightening, and the idea that an animal might have its own opinions and perceptions violates his idea of "dogginess."

As an account of an exchange between two putative experts on the behavior of wolves and dogs, Coppinger's anecdote contains elements of both high drama and low comedy. Coppinger seems aware that wolves will flee when approached, an observation with which we generally concur, although there are exceptions (Jans 2015). The animals in Wolf Park had been in captivity for several generations, which to us suggests that they were well on their way toward becoming domesticated, regardless of their phenotypes. A group bred under human control (presumably selectively) for several generations is undergoing a process of domestication (Morey 2010; Shipman 2011; Pierotti 2012b). Coppinger further admits that he was "reluctant" to approach these animals, suggesting that his body language was probably giving off signals of fearfulness, possibly combined with aggressiveness. Then we have Klinghammer's advice, "Just treat them like dogs," which suggests that he considered these animals dogs and would interact with them as if they were, in fact, his dogs. At this point Coppinger, in his own words, "thumped" an unfamiliar large canid on its side in a way that Klinghammer interpreted as "hitting" her. This is the oddest aspect of Coppinger's story: that an alleged expert on dog behavior would think this action appropriate. Cassi's response is exactly what might be expected from a dog of any breed who has been "hit" by an uneasy stranger. When the wolf responded with aggression, Klinghammer reacted with panic, allowing the situation to further degenerate. Klinghammer should have intervened to stop the aggression. Despite the hyperbole, there is little evidence that Coppinger was in any real danger beyond what existed in his own (and Klinghammer's) mind. He was basically seized rather than attacked as if he were prey. He does not say how he "escaped." We can assume that he simply pulled away from the animals and scrambled out. He describes no actual injuries, other than to his clothing.

This event should be compared to a discussion of how to socialize wolves to unfamiliar adult humans: "With increasing confidence the wolf's fear and uncertainty may morph into aggression. The animal might bite and tug the investigator's clothing. During this period the animal, which is still unsure of itself, gives conflicted signals. Another person should be present outside the enclosure and move forward if the animal becomes more aggressive, which will cause it to stop and withdraw" (Woolpy and Ginsburg 1967, 360). Gordon K. Smith, longtime breeder of wolf-dogs, makes the following statement: "I find that by completely ignoring the wild wolf adult, going into its pen and [acting as if you] do not see it is the best training one can give. . . . One can go in and inspect cubs without attack" (1978, 141). There is no indication that Klinghammer was aware of either of these obvious and straightforward solutions, although both were published decades before the incident at Wolf Park.

Pierotti has had experiences similar to Coppinger's, when wolves seized his arms and pulled at his pants legs. If you stay calm, nothing happens. Holding onto clothing is simply a test or an attempt to control your movements, not a predatory attack. Pierotti has been grabbed by wolves, who will encircle your calf, lower arm, or hand with their mouth, but they do not bite down, and you will not be injured (and your jacket will not require a "tourniquet") unless you jerk away and injure yourself, or your clothing, on their teeth in the process. We cannot emphasize strongly enough that Cassi's action was not a bite, which involves forcefully closing the mouth and jerking the head backward. When wolves grab something to control its movement, of necessity, they use their mouths.

His experience at Wolf Park appeared to have shaped Coppinger's attitude, however, and set him off down the path toward his "garbage dump" model of dog evolution (Coppinger and Coppinger 2001). Because of this single experience, he apparently regards any suggestion that humans could coexist closely with wolves as what he derogatorily refers to as "the Pinocchio approach of taming wolves," although exactly what he means by this statement is unclear; Pinocchio is an inanimate object that comes to life, not an independent creature with its own thoughts and interests concerning the interaction.

THE COPPINGERS' THEORY OF DOMESTIC DOG ORIGINS

It is worth examining the Coppingers' 2001 book, *Dogs: A Startling New Understanding of Canine Origin, Behavior and Evolution,* which has garnered a certain amount of attention in the media (two PBS shows have featured it) and also among those in the scientific community (e.g., Morey 2010; Francis 2015). The belief displayed in the Coppingers' book is that humans control the entire wolf/human interaction, even though there is little supporting evidence for this (see Bekoff 2001). A scholar who has experienced wolves in nature describes the Coppingers' model in this fashion:

> One of these ideas [about how dogs evolved from wolves] is that wolves evolved as scavengers, hanging around human camps. Anyone who has lived in wolf country . . . would find this untenable. Wolves don't scavenge, unless they're starving, or if they opportunistically stumble upon another predator's kill. And, human camps don't usually leave enough meaty refuse to attract a wolf, although they often visit camps out of curiosity. . . . A large mammal carcass would attract a wolf; but humans have been using tools to clean meat from bones for more than [a] million years, and by about 15,000 years ago, when the first unequivocal dogs appear in the archaeological record, humans were very efficient at cleaning up the bones and extracting the

marrow. The leftovers would not feed even one wolf, let alone a pack. Human settlements leave enough refuse; but these only appear about 10,000, just before agriculture. (http://leesnaturenotes.blogspot.com/2010/08/dog-evolution-camp-scavenger-hypothesis.html [posted August 25, 2010])

This last point is particularly salient. Humans did not live in settlements large enough to have dumps that could attract wolves or other, more serious scavengers until the last 10,000 years or so (Morey 2010), whereas there is considerable evidence that the human/wolf relationship began earlier, possibly as early as 100,000 YBP (Vilà et al. 1997; Ovodov et al. 2011; Shipman 2011, 211, 2015; Germonpré, Lázničková-Galetová, and Sablin 2012).

A review of literature on wolf behavior characterizes the Coppingers' idea thus: "A popular hypothesis proposes that the first wolves tamed by humans were those lurking around the campfire hoping to scavenge scraps. With the passing of time and through generations of wolves and humans a few animals eventually lost their fear and became tame. Such a progression is doubtful. Adult wolves are difficult to socialize. The first wolves to interact fully with humans must have been taken from the den early and socialized while still young" (Spotte 2012, 211). This account is overly simplistic and alludes to the "huddling around the campfire in terror" scenario mentioned in our introduction, but it also reveals additional problems with the Coppingers' "garbage dump" model.

The Coppingers argue that wolves self-domesticated, evolving on their own into doglike phenotypes in a process assumed to be analogous to the evolution of the captive silver foxes studied in Siberia (Belyaev 1979; Belyaev and Trut 1982; see also Safina 2015). This Russian study is interesting and may provide insights into how some traits in contemporary forms of domestic dogs appeared, but it sheds little light on the evolution of large socialized wolflike canids found throughout Eurasia or on canids known to North American Indigenous peoples prior to the arrival of Europeans (Fogg, Howe, and Pierotti 2015).

The major reason that the fox model for domestication does not work for wolves is that fox social systems are very different from the family groups in which wolves live. Although they are canids, foxes evolved along a different evolutionary branch than wolves and dogs, and they are not highly social (Francis 2015). They do not form packs or engage in cooperative hunting. As a result, they are not a good model for understanding how social relationships were established between humans and wolves. The fox results demonstrate that it is possible to selectively breed at least one species of canid to become less fearful of humans over only a few generations, and that morphological changes may be associated with behavioral changes (Morey 1994, 2010; Crockford 2006). In contrast, wolves almost certainly continued to look like wolves for thousands of years after they initially established relationships with humans, which calls into question whether results obtained from foxes, or the interpretation of these results by the Coppingers, can be directly applied to wolves.

Several assumptions hidden in the Coppingers' model are violated, even by their own evidence. They assume that there is no selection on these "garbage dump" wolves to retain wolf traits, which would be expected if these animals continued to hunt in addition to scavenging. Many stories from Australia, Asia, and North America describe hunting by the canid companions of the Indigenous peoples. None of these mention dump, or even midden, feeding. The Coppingers further assume that there is no benefit accruing to humans from companionship with wolves. Interestingly, both Safina (2015), who describes complex social dynamics in wolves, and Hare and Woods (2013) seem to fall into the Coppinger model when discussing the transition from wild wolf to domestic dog. Everything in the thinking of these white Americans assumes strong selection to produce a quasi-parasitic animal that is totally dependent upon scavenging from humans, yet no mammal, except possibly Norway rats (*Rattus norvegicus*), has ever evolved to fill such a niche.

The strongest part of the Coppingers' book is chapter 4, "Developmental Environments." Their research demonstrates that one major difference among a variety of modern breeds is that they have incomplete chains of behavior. For example, predatory behavior in wolves involves a sequence of behaviors that occur in a set order, such as *orient/eye/stalk/chase/grab-bite/kill-bite/dissect* (Coppinger and Coppinger 2001, 116). Various breeds lack one or more segments within this chain. Our border collies do everything up to and including the grab-bite and then stop, although, as we suggested above, grabbing with the mouth may not truly involve an actual bite. Experienced trainers manage to control their border collies so that when running sheep they do not use the grab-bite, even though they are fully capable of doing so (McCaig 1991). In contrast, our treeing Walker hound has an exaggerated tracking (stalk/chase) component but lacks the grab-bite in her repertoire; she stops after she has cornered an animal, baying loudly to solicit attention from both humans and other hounds. The Coppingers argue that in their study of feral dogs on Pemba, an island off the east coast of Africa, these dogs lacked all of the predatory motor patterns and had become complete scavengers. In this case, it might have been interesting to see how they responded when exposed to an actual prey item, like a rabbit.

These are interesting observations that allow insight into how different breeds of dog can be trained most effectively and how selection by humans has shaped their behavior. There are breeds, however, in which the entire chain of behaviors is intact, including spitz-type dogs, basenjis, dingoes, and an astonishing variety of medium and large mutts. This grouping includes some of the most wolflike of dog breeds, and their behavior, combined with their appearance, negates all arguments about how dogs have been shaped to be completely different than wolves. These animals can be very wolflike, both in behavior and appearance, and some dogs can survive on their own without resorting to scavenging.

An example is described by the writer Albert Payson Terhune in his 1935 classic, *Real Tales of Real Dogs,* in the story "Aeroplane, the Dog Who Turned Wolf." Aeroplane was a champion border collie, expert at opening latches. Sold to a new owner, he escaped and took up residence under an abandoned pavilion, living by raiding henhouses (he could easily remove the latch) and taking rabbits and wild birds. His condition improved compared with his captive state; he was healthier and stronger. Details of this case are known because Aeroplane was captured after a year on his own and returned to his owner. He resumed living as a "dog," won international championships, and was acclaimed the top border collie in Canada. This is similar to stories concerning dingoes and wolves raised by humans who reverted to the wild but then returned to hunt with their human companions.

Ray Coppinger's work on Pemba represents a deliberate attempt to find a population of dogs living under conditions that mimicked (from his perspective) the conditions of the Mesolithic (Coppinger and Coppinger 2001). Several issues concerning this population, however, suggest it may not be representative of ancestral dogs. To begin with,

Pemba has an entirely Islamic human population. Islam has a tradition of regarding dogs as unclean (Foltz 2006). Therefore, when Coppinger asks Pembans if they pet their dogs, he should not be surprised to find out that they never have. These dogs truly are scavengers, and they do seem to have achieved a sort of equilibrium, evolving toward a relatively consistent phenotype. The issue is, however, whether there is evidence that these dogs represent the ancestral state of modern dogs or simply one of many possible outcomes. They bear no resemblance to dogs kept by Indigenous peoples in many parts of the world. Weighing around seventeen kilos (thirty-five pounds), they scavenge primarily because there is little else to eat on Pemba. In contrast, the early dog/wolves had a wide range of potential prey and could choose to hunt.

The Coppingers' book provides information about only a few breeds, or categories, of dog. It supplies no data or insights into the evolutionary history of wolves and their interactions with humans, even though these are clearly significant aspects of the histories of many cultures around the world (Pierotti 2011a; Fogg, Howe, and Pierotti 2015). The Coppingers write as if all traits found in dogs evolved after they became separated from wolves, even while acknowledging that many wolf behavioral traits can also be found in dogs. Even traits like high red blood cell counts, allowing greater endurance in many dog breeds used as beasts of burden (such as for pulling sleds), are treated as if they evolved in dogs completely independently of wolves, even though such characteristics would clearly be favored in wolves involved in long-distance pursuits.

This approach leads down some confusing pathways. The Coppingers' discussion of sled dogs deals only with how to maximize running speeds for racing competitions, because Ray Coppinger competed in sled dog races. Sled dogs were originally bred not for speed but for the strength to pull loads, which is why malamutes and huskies have stronger shoulders and broader chests than wolves, and in consequence are slower runners. Such dogs were not bred for the sporting pleasure of people of European ancestry.

In fact, the Coppingers provide no discussion of any functions of dogs that aren't exclusively of concern to people of European ancestry. This is another example of Euro-bias, seen most obviously in their argument that the human/wolf dynamic is not mutualistic. The illogic of this conclusion derives from their observation that contemporary humans could survive without dogs; therefore dogs should be considered para-sites, or amensals. By contrast, the argument we make, supported by evidence provided by Indigenous peoples, is that, upon their arrival in Europe and North America, humans might not have survived without cooperative hunting and food sharing with wolves (Schleidt and Shalter 2003; Shipman 2015).

To understand the Coppingers' thinking, it is essential to read their chapter on household dogs. They describe this relationship as *dulosis,* or "slave-making," because dogs are "forced" to work for humans (2001, 255). They refer to dingoes as "crop pests" (280), a pejorative term appar-ently meant to suggest that dingoes can survive only by scavenging off human activities, a proposition that would offend most Australian Aboriginal people (Rose 1996, 2000, 2011; chapter 6), not to mention that Aboriginals did not have crops as such; therefore this insult is cast totally within a colonial mind-set.

The Coppingers argue that dogs represent a huge drain on the economy and the ecosystem because they require large expenditures on food and care. This line of thinking misses the essence of the relationship—many humans value their dogs as they value other humans, as companions that support you, protect you, provide comfort, and can even catch food for you. This is understood by trappers and herders in central Siberia (chapter 5), Indigenous peoples of the Americas (chapters 7–8), and Australian Aboriginals (chapter 6) as well as by untold numbers of people in Europe and North America who value their canid companions.

This failure to understand, or even acknowledge, the strong bonds between humans and their canid companions helps explain why the Coppingers think dogs evolved from wolves that depended upon refuse. Oddly enough, despite his recognition of complex social interactions in wolves, Safina (2015) argues that wolves fed on human scraps. Similarly, Hare and Woods create an elaborate argument outlining that humans were competitive rather than cooperative with wolves (2013, 29). They also state that wolves "do not cooperate with humans and are relatively uninterested in humans, even if raised by people" (178). Hare and Woods cannot imagine a reciprocal relationship between *Homo sapiens* and *Canis lupus* that coevolved from a cooperative foraging relationship. The Coppingers recognize potential benefits in the human/dog relationship to dogs but not to humans. Reading the Coppingers' book, one gets the impression they don't really like dogs; they continually refer to the "flaws" of dogs and posit that "gadgets" could easily replace dogs—a questionable assertion, given that dogs possess a sense of smell that has not yet, and may never be, matched by machines (Morey 2010).

The reasons for R. Coppinger's attitude become more apparent when you read his latest work (Coppinger and Feinstein 2015). Here it is revealed that their sled dog work involved some 4,000 individual dogs, and their work on herding and flock-guarding dogs involved an additional 1,500. These are puppy-mill numbers—it is hard to develop relationships when you have so many dogs; individual personalities become blurred (Pierotti 2011a).

To emphasize this point, this latest work by Coppinger and Feinstein is one long argument about why nonhuman animals should be considered "machines," an idea derived from Cartesian thinking. They define behavior as "movement in space and time"; under such criteria, it can be argued that airplanes and automobiles are "behaving." They also argue that dogs are incapable of complex mental states and cannot experience emotions. This recent work reveals how little respect R. Coppinger has for dogs as living beings (see also Bekoff 2001).

Chapter 10 in the Coppingers' 2001 book deals with the use of DNA to assess canid systematics, with specific attention paid to Vilà et al. 1997, the only major paper on *Canis* DNA published before their book was written. The Vilà et al. work troubles the Coppingers, primarily because that study suggests that the separation between dogs and wolves may go back more than 100,000 YBP, which creates an obvious conflict with the Coppingers' notion that dogs did not appear until humans had permanent settlements, around 12,000 YBP.

The Coppingers are at odds with evolutionary theory and modern systematic approaches, seeming to assume that "tameness" is genetically driven rather than a conditioned response to social environment. We have already mentioned the failure to consider selection pressure on animals to remain wolves, even after wolves have supposedly taken up scavenging, which represents only one alternative among many possible methods of foraging. The Coppingers write as if all selection was directed only toward traits that allowed these ancestral dogs to coexist as parasites with humans. In reality, traits do not evolve independently of other aspects of the phenotype; some traits are simply the result of correlation to other traits or of random developmental noise. This is a point made by Gould and Lewontin in their 1979 "The Spandrels of San Marcos," which deconstructed the sort of the simplistic adaptationist paradigm exemplified by the Coppingers' work.

The Coppingers chose only ideas that conform to their way of thinking. They state, "Wolves, dogs, coyotes, and jackals can all interbreed, or potentially interbreed, and produce viable offspring. Therefore, they are not 'good' species according to Mayr's definition" (the Biological Species Concept, or BSC) (2001, 276). This reveals a very naïve position on the nature of species in the twenty-first century and converges on a view held by people who oppose conservation. Such an argument shows ignorance of recent thinking about the role of hybridization in evolution and of contemporary ways of thinking about species formation.

Ray Coppinger explored this idea in more detail in an intriguing but unsatisfying work that opens with the statement, "Species can be considered as moving targets" (Coppinger, Spector, and Miller 2010, 41). This starts the argument off immediately on the wrong pathway and renders their results confusing. Environmental conditions are the constantly moving targets; species are merely the troubled marksmen trying to keep up. This phenomenon has been described by Van Valen 1973 as the Red Queen hypothesis, which assumes that organisms must constantly adapt, evolve, and proliferate not merely to gain reproductive advantage, but also simply to survive while pitted against ever-evolving opposing organisms in an ever-changing environment (Bell 1982).

By identifying the species, rather than the environment, as the moving target, Coppinger, Spector, and Miller (2010) reveal another form of Euro-bias: the idea that species change whereas the environment stays constant—that is, the "balance of nature" (Pierotti 2011a). Coppinger, Spector, and Miller identify a potentially serious issue in the way canid evolution is studied—the idea that there is only a single origin of domestic dogs, and the suggestion that once new forms appear they should be reproductively isolated from their ancestral forms (we discuss this further below). They point out, correctly, that many people charged with the conservation of endangered species have very naïve ideas of what a species is as well as of the most effective ways of protecting one, especially if hybridization is occurring between an endangered species and a nonendangered relative. This reveals flaws in the Endangered Species Act and the Biological Species Concept (O'Brien and Mayr 1991; Pierotti and Annett 1993).

What Coppinger, Spector, and Miller (2010) fail to realize, however, is that hybridization within the genus *Canis* is logical and predictable. "The pattern of occurrence of naturally occurring intrageneric hybrids in [vertebrates] is non-random with regard to mating system, and may in fact result from adaptive behavior with evolutionary consequences" (Pierotti and Annett 1993, 670). In vertebrate lineages in which females choose males on the basis of their quality as parents and providers, sometimes the best mate available is a member of another, closely related species (Good et al. 2000). Species with male parental care (wolves are one of the best examples of this among mammals) show more frequent hybridization and reduced development of reproductive isolation (Pierotti and Annett 1993), whereas complete reproductive isolation is the basis of the BSC.

The Coppingers may be correct in their assertion that "if there is recent hybridization [between wolves and dogs], it implies that hybridization has taken place from the very beginning of dogs" (Coppinger and Coppinger 2001, 289). Pierotti has argued: "Evolution of dogs is best imagined as a tapestry rather than a phylogenetic tree, resulting from multiple domestication events between humans and wolves beginning at least 40,000 YBP and continuing until this day in some areas. These separate domestication events resulted in subsequent crossing among lineages. . . . The history of what we call dogs is completely interwoven, with various wolves appearing as a sort of golden thread providing overall resilience and genetic diversity within many lines" (Pierotti 2014). Figure 1.2.10 in Coppinger, Spector, and Miller 2010 presents an example of such a "tapestry," in which reticulate evolution results in frequent introgression within the genus *Canis*. What they fail to recognize is that this interweaving of lineages almost certainly results from environmental disturbance, including efforts at extermination by humans of European heritage (McIntyre 1995). Before Europeans invaded North America, gray wolves, coyotes, red wolves (*Canis rufus*), and tweed wolves (*Canis lycaon*), were distinct, reproductively isolated lineages that probably conformed to the Biological Species Concept. Contemporary American *Canis* present the most complex hybridization pattern of all mammals, which resulted from heavy persecution of all types of wolves over the last 150 years, with consequent extirpation from much of their former range. Male red and gray wolves dispersing from their families and seeking mates are likely to encounter more coyotes than conspecifics.

Basing your understanding of North American canid species upon Mayr's BSC might be a tenable intellectual stance if wolves lived in an unchanging world. Like others arguing that domestic dogs are a species, the Coppingers fail to address that if this is true, what are the characteristics by which this species can be identified? This is not a trivial question: as emphasized above, we are discussing a grouping that includes forms as diverse as Chihuahuas, toy poodles, cocker spaniels, bloodhounds, Great Danes, bull mastiffs, Siberian laiki, dingoes, malamutes, and Eskimo dogs. This especially ignores breeds such as Saarloos wolf-dogs Czechoslovakian wolf-dogs, utonagans, and even German shepherds, which have specifically been bred using wolves in their pedigrees.

In their chapter "What's in the Name *Canis familiaris*?" the Coppingers (2001) discuss how scientific names are established. They leave out, however, any mention of the criteria by which such names are assigned or subsequently changed as our understanding of the evolutionary history of a group advances. To remove all doubt within the scientific community about which member of the genus *Canis* is ancestral to domestic dogs, it was correctly decided that wolves and dogs should be considered to be members of the same species. The Coppingers fail to present an accurate reflection of the situation: where the process was as described above in that all synonyms that ever existed for *Canis lupus* were listed in recent systematic revisions. According to current systematic thinking, domestic dogs are currently considered simply to be forms of wolves. The Coppingers' complaint should be directed at the tradition in modern systematics, which is to assume that species (and subspecies) have only a single evolutionary origin.

To make things even more confusing, the Coppingers identify J. H. Honacki, lead editor of the 1982 volume *Mammal Species of the World: A Taxonomic and Geographic Reference,* as the individual who changed the scientific name of domestic dogs from *Canis familiaris* to *Canis lupus familiaris*. It is unlikely that as lead editor, Honacki actually had anything to do with this specific decision (the late Chris Wozencraft, who wrote the carnivore section of D. E. Wilson and Reeder 1993, came to a similar conclusion).

This point relates to the issue of interbreeding among lineages (hybridization). As dogs were derived from wolves, it seems likely that the process of domestication occurred independently in different parts of the world at different times (Morey 1994, 2010; Tsuda et al. 1997; Vilà et al. 1997; Leonard et al. 2002; Savolainen et al. 2002; Derr 2011; Shipman 2011; Thalmann et al. 2013; Pierotti 2014). Having multiple origins suggests that various types of dogs would initially be descendants of separate wolf populations. Thus, if they ended up crossing with each other centuries or even millennia after the initial events, their offspring would be the result of *reticulate evolution,* in which an evolutionary line has more than one ancestor (e.g., Frantz et al. 2016). The Coppingers' basic point is likely to be true; the crucial consequence, however, is that when reticulate evolution occurs it is inappropriate to consider the resulting forms to be species, or even subspecies (Pierotti 2012a, 2014), as the Coppingers argue.

DNA, DOMESTICITY, AND CONFUSION

This complex hybridization pattern found among American wild canids presents numerous problems for investigators looking into the usefulness of DNA sequencing as a tool for understanding evolutionary patterns within the genus *Canis* (see Coppinger, Spector, and Miller 2010). Many scientists assumed that once we had complete DNA sequences from organisms, we would be able to resolve all sorts of questions concerning species identity and evolutionary lineages, such as what is a dog compared with a wolf? For the past quarter century numerous individuals throughout Europe and North America have been working on this issue, but we have little more agreement concerning evolutionary relationships among breeds of dog than we did when this work was initiated. What these DNA results have established, however, is that our assumptions, even about issues that initially seemed as clear-cut as the differences between wolves and coyotes, often have complex histories.

Coppinger, Spector, and Miller 2010 take issue with the results reported in some of the earliest efforts in this area (e.g., Vilà et al. 1997). Their main problem apparently involves timing of the possible "split" between wolves and dogs, because Vilà et al. use a "clock" based on rates of mutation in mitochondrial DNA (mtDNA) to suggest that this split may have occurred more than 100,000 YBP. Such "clocks" assume a fixed mutation rate in mtDNA, which is buffered from selection by its location within cells. Vilà et al. chose an event whose timing is reasonably well established: the separation of wolves and coyotes (*Canis latrans*). By comparing the difference in DNA sequence between wolves and coyotes and subsequently between wolves and various types of dogs, they provided an estimate of the timing of this latter separation.

In an effort to refute this estimate, Ray Coppinger turned to Lynn Miller and "talked Lynn into teaching him how to read the [Vilà et al.] paper" (Coppinger and Coppinger 2001). Miller's analysis focused on the timing question rather than the major results of Vilà et al., which was the first significant work to (1) confirm that dogs have a polyphyletic lineage, and (2) clearly establish that wolves are the sole ancestors of domestic dogs. If Coppinger (and Miller) had understood this, they might not question why dogs are not a species but instead should be considered an assemblage of related organisms whose ancestors were also an assemblage of ancient wolves, each of which probably contributed different parts of what has become known as the "dog genome" (Pierotti 2012b, 2014).

We concur with the Coppingers on one point, which is that Vilà et al. (1997) are scrupulous in avoiding any sort of scientific name for their study species. Neither *Canis lupus, Canis familiaris,* nor any variant thereof can be found in this publication, which we believe represents the first time that the journal *Science* published a major biological paper without requiring the authors to identify their study organisms using scientific nomenclature. This practice has now become typical of all papers *Science* publishes on "dog evolution," which we regard as tacit recognition of the complex evolutionary history of domestic dogs (Pierotti 2014). We part from the Coppingers over their contention that dogs should be a separate species, or alternatively that dogs, wolves, coyotes, and jackals should all be considered members of the same species because they are capable of interbreeding.

This reluctance to provide a scientific name reflects uneasiness on the part of Vilà et al. 1997 (as well as subsequent papers on this topic) concerning whether identifying wolves and dogs as all being the same species would lead to attempts to delist wolves from endangered species status. This is a legitimate concern; an earlier publication (Wayne and Jenks 1991), which argued that red wolves (*Canis rufus*) appeared to be a hybrid between wolves and coyotes, generated attempts to delist *C. rufus* based on the logic that hybrids are not protected under the Endangered Species Act (Coppinger, Spector, and Miller 2010; see also Pierotti and Annett 1993 for discussion of this theme). This misuse of Mayr's BSC by right-wing opponents of species conservation generated controversy about how genetic evidence should be used in determining species status, and whether the BSC functioned as an impediment to effective conservation efforts (O'Brien and Mayr 1991).

Problems that arise from the competing, and apparently incompatible, views presented by the Coppingers and their collaborators and the UCLA research group (Robert Wayne's lab at UCLA produced numerous canid genetic papers, e.g., Vilà et al. 1997; vonHoldt et al. 2010) derive from different ways of doing science. The Coppingers look primarily at behavior and only superficially at anatomy, while ignoring evolutionary processes. In contrast, the UCLA group looks primarily at DNA sequences and largely ignores behavior and anatomy. Both groups ignore biogeography and ecology. Canid scholar Mark Derr has commented about DNA evidence: "Rapid advances in genome sequencing in recent years have focused so much attention on the biology of domestication that the equally important cultural influences have been overlooked . . . [because of] convincing genetic evidence that most domestication [events] occurred in more than one place at one time . . . sometimes . . . at distant locations nearly simultaneously" (2011, 51). Failure to consider cultural factors further reveals a logical split between geneticists and biologically oriented anthropologists like Hunn (2013), Morey (1994, 2010), Crockford (2006), and Shipman (2011, 2015), all of whom emphasize the role of humans and wolves in shaping one another.

Biogeography and ecology, both of which are strongly linked to cultural factors, are crucial to understanding relationships between humans and wolves. Modern humans evolved in sub-Saharan Africa, where there are no wolves. Humans seem never to have established close commensal relationships with jackals, or coyotes for that matter, which are nonpack-forming canids not readily socializeable within a group context. Once humans moved out of Africa into Asia, and eventually into Europe and North America, they encountered true wolves, and within a short period of time the relationships we discuss were initiated.

Genetic studies of dog ancestry certainly don't all agree with one another; however, they do consistently demonstrate that it is likely that multiple origins, both temporally and spatially, characterize dogs and that interbreeding, both among various types of dogs and with wild ancestors, has created a complex and rich ancestry (Verginelli et al. 2005; Ovodov et al. 2011; Shipman 2011, 2015; Pierotti 2014). The striking phenotypic and genetic diversity of dogs clearly indicates that their founders were recruited from a large and varied wolf population (Verginelli et al. 2005). Phylogenetic trees of dog and wolf hypervariable mtDNA sequences show that dogs group into several clades (Tsuda et al. 1997; Vilà et al. 1997; Leonard et al. 2002; Savolainen et al. 2002; Frantz et al. 2016), consistent with independent origins from multiple wolf matrilines.

Over the last twenty years, depending upon which DNA-based study is cited, there have been at least four (possibly six) separate "origins" of domestic dogs reported: eastern Asia (Savolainen et al. 2002, Niskanen et al. 2013), the Levant (Leonard et al. 2002, vonHoldt et al. 2010), southern Asia (Skoglund, Götherström, and Jakobsson 2011); and most recently Europe (Thalmann et al. 2013), each presented as if it were the "original" and "only," site where this process occurred. These analyses ignore the significant crossbreeding that has occurred over the millennia since establishment of the human-canid relationship in the Paleolithic (T. N. Anderson et al. 2009; Coppinger, Spector, and Miller 2010; Pierotti 2014). There are obvious weaknesses in these other studies; for example, a widely cited work (vonHoldt et al. 2010) does not include Native American dogs, Alaskan huskies, or Siberian laiki in its analyses, although these are among the most wolflike of "dog" breeds (figure 1.1; Pierotti 2012a, 2012b). In a similar fashion, Thalmann et al. 2013 leaves out "specimens from the Middle East or China, two proposed centers of origin" because "no ancient dog remains older than 13,000 years are known from these regions" (873).

These omissions would not be crucial if authors acknowledged the existence of multiple origin sites where wolves transformed into early domestic forms. One of the most recent publications argues that "none of the wolf lineages from the hypothesized domestication centers is supported as the source lineage for dogs, and that dogs and wolves diverged 11,000–16,000 years ago in a process involving extensive admixture and that was followed by a bottleneck in wolves." The authors further state, "At the beginning of the domestication process, dogs may have been characterized by a more carnivorous diet than their modern day

FIGURE 1.1 One phenotype of Alaskan husky, a breed that has
resulted from crossing various northern breeds of dog. (Photograph
by Asigglin, Wikipedia)

counterparts, a diet held in common with early hunter-gatherers"
(Freedman et al. 2014, 2). This brief sampling reveals considerable debate
concerning both the time and location of canid domestication. The most
recent publication on this topic (Frantz et al. 2016) has finally acknowledged the multiple origins of domestic dogs, although a critical reading of
all of these studies has been pointing in this direction for many years
(Pierotti 2014). This recent work by Frantz et al. suggests that all of these
sites represent independent domestication events, and that derivation of
domestic canids from various types of wolves did not happen once, either
13,000 or 35,000 years in the past, but probably continued into historic
times in some parts of the world (Losey et al. 2013; Pierotti 2012a, 2012b,
2014; Fogg, Howe, and Pierotti 2015). "Domestication of the dog took
place multiple times in many places involving any number of wolf
lineages" (Spotte 2012, 20).

Looking at ancient mtDNA, Verginelli et al. 2005 found that Italian
subfossil canids compared to a worldwide sample of 547 purebred dogs and
341 wolves revealed highly diverse sequences that joined three major clades
of extant dog sequences, indicative of polyphyletic origin. Phylogenetic
investigation highlighted relationships between ancient sequences and
geographically widespread extant dog matrilines as well as between ancient
sequences and extant wolf matrilines of east European origin. These results

provide support for the involvement of European wolves in the origins of the three major dog clades and also suggest multiple independent domestication events (see also Thalmann et al. 2013; Frantz et al. 2016). Discovery of a 33,000-year old probable domestic canid in the Altai Mountains of southern Siberia (Ovodov et al. 2011), combined with other domestic dog remains of roughly the same age recovered from a cave in Belgium (Germonpré et al. 2009) and extensive evidence from east and south Asia further suggests that the domestication of dogs was unlikely to have happened at only one location.

One obvious finding is that evidence from archaeology and genetics combines to contradict the model proposed by the Coppingers (2001) of a single domestication event that took place within the last 12,000 years. Results presented by vonHoldt et al. 2010 show at least six separate events leading to various forms of "ancient" dog: (1) basenji; (2) an east Asian group, including akita, chow, shar pei, malamute, husky, and possibly including dingoes; (3) a line leading to salukis and Afghans; (4) a line leading to American Eskimo dogs and Samoyeds; (5) a line leading to Kuvasz and Ibizan hounds; and (6) a line that seems to lead to most extant dog breeds.

In addition, many of the spitz-type "dog breeds" commonly crossed with wolves to produce supposed "wolf hybrids" or "wolf-dogs" are more closely related to wolves than to what most people consider "dogs." For example, the German shepherd, a breed with a somewhat wolflike appearance derived from a working-dog lineage, appears to be most closely related to schnauzers, according to vonHoldt et al. 2010. Breed founders, however, indicated that they used dogs with recent wolf ancestry to establish the breed (see chapter 9). Wolf-dog breeder Gordon Smith (1978) claims that some German shepherd breeders switched pedigrees between tamed wolves with AKC-registered shepherd studs to produce better-looking "German shepherds." Interestingly, the recent publication on the multiple origins of dogs (Frantz et al. 2016) places German shepherds with other breeds known to involve crossing with wolves, such as the Czech wolf-dog, and among ancient spitz breeds. This suggests that vonHoldt et al.'s (2010) placing of German shepherds with working dogs such as schnauzers and Portuguese water dogs is likely to be incorrect. At the very least, both of these DNA-based results cannot be correct. Frantz et al. (2016) are the first investigators looking at DNA evidence to explain dog evolution to have included DNA from Czech and Saarloos wolf-dogs. No one has examined DNA from central Asian laiki, North American Indian dogs, or any of the Belgian shepherd (sheepdog) breeds, which are among the most wolflike in appearance of recognized breeds.

In the end, however, culture may be more important to understanding what a *dog* represents to humans than DNA, because all cultural traditions, and maybe all individual humans, have specific concepts of what a dog is—a member of the genus *Canis* that fits their particular way of life. To one of our neighbors in rural Kansas, the dog is a treeing Walker hound, which fits his personal obsession with nighttime raccoon hunting. To another neighbor, the dog is a Labrador retriever as a companion in daily life and a loyal partner in bird hunting. To our family members, dogs are small and cannot live outdoors. All these dogs are clichés at some levels, but they represent what image comes into people's minds when they think the concept of *dog*.

This becomes more important in terms of the larger evolutionary question: what was the "canid companion" image in the minds of humans who first established relationships with wolves? We can see some hints in the animals chosen by the various cultural traditions we discuss in other parts of this book. To the hunters of Siberia and their modern descendants, dogs are barely domesticated animals that would be mistaken for wolves by the vast majority of Americans. To Native Americans, dogs were wolves; in fact, in many cases they did not make any distinction because to them it probably did not matter (Fogg, Howe, and Pierotti 2015; chapter 7). To Australian Aboriginals, the dingo is their image of the canid with whom they shared their cultural development, and they seem unsure of how to deal with the "domestic dogs" that arrived when their land was invaded by Europeans in the eighteenth century (Rose 2000, 2011; chapter 6). Each cultural tradition developed with specific images of the canids suited to share its particular way of life. This is why it is so hard to define the term *dog*: all traditions, and even specific individuals with their own personal cultural traditions, have very different images of what type of canid is best to share their life and living space.

CHAPTER TWO

Cooperation between Species

As far as raven was concerned, Man, the new predator, was probably just a surrogate wolf who also usually hunted in packs.

Bernd Heinrich, *The Mind of the Raven*

At least since Darwin, a scientific cultural tradition has existed that assumes competition is more important than cooperation in structuring ecological communities. Part of this tradition is the assumption that humans must have controlled the initial interaction between humans and wolves, and that it took thousands of years and an elaborate breeding program to transform wolves into dogs (Coppinger and Coppinger 2001; Francis 2015).

As an example of how these themes interweave, the Coppingers created an argument about human/wolf interactions and the origin of dogs that posited two possibilities, either (1) dogs and wolves must be different species (oversplitting from a taxonomic perspective), or (2) all members of the genus *Canis* should be considered members of the same species (overlumping from a taxonomic perspective) (Coppinger and Coppinger 2001, 273–82). Following a similar logical path, they and Francis (2015) assert that humans and wolves are not cooperative but competitive with each other, which aligns them conceptually with anti-conservationist politicians trying to remove endangered species status; they also argue that contemporary wolves are competitors with humans, for example, with regard to "game" animals, such as moose, caribou, elk, and deer (Haber 2013; Jans 2015).

In contrast, more modern scientific approaches explore how different species can create and live within physiologically and behaviorally cooperative relationships that benefit both species (Dugatkin 1997). In recent years, Western scientists have realized that cooperation is more important in both nonhuman and human societies than competition (Worster 1993, 1994; Dugatkin 1997; Pierotti 2011a). Among nonhuman species, it turns out that 85–95 percent of behavior is affiliative or cooperative rather than aggressive or competitive (Shouse 2003). Cooperation and "facilitation" are found not only within species but among groups of species within ecological communities.

EVOLUTION OF COOPERATIVE RELATIONSHIPS

The most indisputable proof of cooperative relationships between humans and other forms of biological life is that, like every other multicellular life form, we contain mitochondria within our cells that are genetically distinct from "us" (what most people consider "our own genes" consist of the DNA found only within the nuclei of our cells, not extranuclear DNA). Mitochondria were initially described as "organelles" that evolved inside of eukaryotic cells, but in actuality they are self-reproducing distinct organisms with their own genetic heritage that live symbiotically inside our cells. This explains why mitochondrial DNA is passed only through the maternal line: egg cells contain mitochondria, whereas sperm do not donate mitochondria to the embryo. Mitochondria provide energy to host cells through the process of aerobic respiration; they are characterized as the "powerhouses" of cells because they provide that energy in exchange for being able to live in a nutrient-rich protective environment offered by their hosts (E. C. Nowack and Melkonian 2010). Their cellular environment buffers them against natural selection, so that supposed neutral mutations accumulate over time through random processes, generating a sort of "molecular clock." These two types of interacting cells (mitochondria and their hosts) shaped each other through a process called *endosymbiosis*. Without this relationship, there would be no plants, animals, or other multicellular life-forms (Margulis 1998; McFall-Ngai et al. 2013). Multicellular (eukaryotic) cells result from a merger between different life-forms rather than being a special "creation" event, and their existence "doesn't require anything more complex than living together in peace and harmony" (Prothero 2007, 154).

Most people, however, think of cooperation as taking place between multicellular (eukaryotic) species, such as plant/pollinator relationships in which the pollinators, whether they are birds, insects, or even mammals, are visible complex organisms easily recognized by humans. There are numerous examples of different species of birds or fishes sharing a nest, with larger species providing protection against predation to smaller species, which in turn act as sentries for the larger species (see Walters, Annett, and Siegwarth 2000). This is the sense we will employ when discussing the evolution of cooperation between humans and wolves, especially the phenomenon of cooperative hunting.

COOPERATIVE HUNTING

Cooperative hunting, as demonstrated by an increase in successful prey capture observed when two or more individual animals engage in a hunt, is a widespread phenomenon (Packer and Ruttan 1988). In many cooperatively hunting species, searches for food items have been described as opportunistic, because individual predators are assumed to be conducting simultaneous individual hunts, with each animal trying to maximize the probability of catching the prey for itself. True coordination is assumed to exist only if individuals play different roles, which implies that at least some individuals adopt roles with a lower probability of personal success or a higher risk of injury—for example, hunts in which some individuals act as chasers while others block escape routes, as in hunting practices of lions or African wild dogs. Such coordination is known in only a handful of species, primarily mammals and birds, although some schooling fishes forage in this manner (Major 1978). Individual role specialization within coordinated hunts is even more rare and has been observed in only two published studies to date (Clode 2002; Bshary et al. 2006). Communication between group members to initiate a coordinated search for suitable prey (intentional hunting) is known only in one population of chimpanzees (Boesch 1994).

Lack of examples on this topic might reflect bias on the part of ecologists trained in the Western scientific tradition, which assumes that competition is the dominant form of interaction, rather than indicate an absence of such cooperative interactions. Pierotti had an opportunity to observe, courtesy of the family pets, a mixed-species cooperative hunting group in which individuals assumed different roles. The participants were three domestic cats (one large orange male and two smaller females, one a Maine coon mix) and the family dog (Hungarian puli/poodle). I (I am switching to first person for the narrative) lived on a one-acre lot in a mixed oak-conifer forest near the Santa Cruz Mountain town of Felton, California. The plot contained large oak trees, one of which was well separated from the others and had several large branches that overhung the roof. Western gray squirrels, *Sciurus griseus,* were present in the area, although we rarely saw them in our yard. The reason became obvious one day when I observed a gray squirrel running across the roof, pursued by the coon cat. The squirrel leaped into the oak tree, and the cat ceased pursuit. What happened next was unexpected. The smallest cat, a black and white female, came across the roof and followed the squirrel into the tree. The squirrel went out onto a narrow limb, but the cat followed. I expected the squirrel to jump to the ground, but then I realized that the large male cat and the dog were both waiting below the tree about ten meters apart, each blocking an easy avenue of escape. All the animals were silent, including the squirrel, which moved farther out on the limb as the small cat climbed closer. The coon cat remained on the roof, blocking any attempt to escape in that direction. This situation lasted for more than thirty minutes, with the squirrel hanging out at the very tip of the branch. The small cat was within a meter of the squirrel but seemed unwilling to proceed further along the slender limb. At this point, I came to the rescue, distracting the predatory group and allowing the squirrel to climb high into the tree.

This example involved four unrelated individuals of two species hunting cooperatively, with each assuming a distinct role. These individuals often operated independently and did not usually move around as a group, but they were capable of employing their unique skill sets in a cooperative manner. The coon cat was the primary pursuer; she was fast, athletic, and in her physical prime. She did not follow the squirrel into the tree, allowing the smaller, more agile female cat to assume this role. The dog and the large male cat, who could not climb well and never went onto the roof, assumed roles where they could go after the squirrel if it tried to escape on the ground.

It is unlikely that this was an isolated event. These four domestic animals had free run of our land at all times, and we often found the remains of squirrels that had been eaten, so it is probable that this cooperation frequently led to successful hunts.

The relationship these individuals had was being part of the same human-assembled community. Domestic cats are rarely described as cooperative hunters, and dogs and cats are assumed to be "enemies," although this clearly is not the situation if they are raised together. The cats and dog lived together as a group for more than ten years, spent considerable time together, and got along well. The dog, the oldest individual, established a congenial relationship with each cat as it joined the household. The male cat and dog often played together. When the dog began to senesce at the age of fifteen, the two female cats (the male cat had died previously), spent hours huddled up against him, grooming him with their tongues.

Another cooperative relationship originally considered to involve competition comes from Native American accounts of badgers and coyotes indicating they were "friends" and hunted together (several examples can be found in Voth 2008). Western ecological concepts, shaped by competition-driven ideas of community dynamics, categorized the relationship between coyotes and badgers as competition between predators (Minta, Minta, and Lott 1992). Assumed competition was emphasized in a documentary, *Badgers: Dishing the Dirt* (*Profiles of Nature*, April 23, 2005, and regularly repeated), that included faked footage implying that badgers and coyotes are rivals, although the two species are rarely shown actually interacting. Empirical study reveals that these two species are cooperative, at least at times. Coyotes and badgers spend a lot of time wandering around together. When they see a ground squirrel, the coyote gives chase. If the squirrel enters a burrow, the badger will dig up the burrow, or it and the coyote will dig together. If the squirrel stays put, badger often gets it. If the squirrel attempts to escape by using another burrow exit, the coyote often gets it. Coyote and badger catch more squirrels when they hunt together than when they hunt alone (Minta, Minta, and Lott 1992). Film footage of this cooperative relationship can be seen in a documentary produced by National Geographic in 1995, *Yellowstone: Realm of the Coyote*.

Both examples show that members of the genus *Canis* are obviously capable of cooperating with other carnivore species to increase hunting success. This proclivity of canids to cooperate seems to be a family characteristic. I (Pierotti) currently have a scent hound and a border collie, a herding dog, two breeds from very different genetic lines within the grouping "domestic dog" (see figure 1 in vonHoldt et al. 2010). Both first appeared as strays. It was apparent that the hound had been living on her own for some time; she was in good shape but had many scratches and was heavily infested with ticks. My land includes a full acre inside a six-foot chain-link fence, and woe betide the raccoon, possum, squirrel, or even coyote that crosses that fence line. The hound corners the "varmint," baying loudly, typical behavior for this breed, which is known as a "coon hound." After the hound recruited the border collie as a partner, the hunting pattern changed: the border collie circles around behind, dashing, lunging, and biting at the luckless prey, showing the *eye/stalk/chase/grab-bite* behavioral sequence (Coppinger and Coppinger 2001). On some occasions the border collie has made a kill (the hound does not grab-bite or kill-bite). On excursions around prairie grasslands surrounding my house, the cooperation of the two is even more apparent. While on her own, the hound apparently supported herself by digging out prairie voles (*Microtus ochrogaster*) and cotton rats (*Sigmodon hispidus*). When she finds a burrow complex she uses her nose to locate active burrows and begins to dig. (She will kill and eat small prey up to the size of eastern cottontail rabbits, *Sylvilagus floridanus*.) The border collie digs as well, finding a hole close to and opposite from where the hound is digging. The border collie lacks the patience and determination of the hound, assuming a role strikingly similar to that of the coyote in the coyote/badger interaction: she watches other possible escape holes, and when a rodent makes a break she often captures and consumes it. Her presence makes the burrowing rodents less inclined to attempt escape, increasing the success rate of the hound as well.

Interestingly, cooperation between coyote and badger and between Walker hound and border collie is similar to recent observations from the Red Sea on the highly coordinated and communicative interspecific hunting between groupers, *Plectropomus pessuliferus*, and giant moray eels, *Gymnothorax javanicus*. In these fish, associations are nonrandom; individual groupers signal to individual moray eels to initiate joint searches. If the grouper recruits an eel, it leads it to a specific portion of the reef, indicating hiding places of prey by pointing with its snout. This signaling appears dependent on grouper hunger level. Both partners benefit from the association. Eels enter coral heads where the thicker-bodied groupers cannot go, creating a situation whereby prey fish must choose either to remain in the complex habitat of the coral, where they are vulnerable to the eel, or to make a break for it, exposing them to the grouper. Benefits of this joint hunting result from complementary hunting skills, reflecting the evolved strategies of each species, rather than individual role specialization during joint hunts. As with the border collie and the hound, the foraging partner that catches a prey item immediately swallows it whole, making sharing of a carcass impossible (Bshary et al. 2006). The hound and border collie do share rabbit kills, however, pulling the carcass apart, each consuming her portion.

Cooperative foraging emerges when multiple individuals are more successful than individuals hunting alone. This is especially true when individuals have different skill sets or physiognomies that are complementary. This is clearly the case when organisms from groups as different as birds and mammals cooperate, as in foraging associations at sea between marine birds and mammals (Pierotti 1988a, 1988b). Smaller marine mammals, like sea lions and dolphins, and even humpback whales (*Megaptera novaeangliae*), feed in groups, especially when going after huge aggregations of schooling fish, which are very effective at confusing and evading attacks by solitary predators. The ocean is a three-dimensional environment where prey can escape in every direction except through the surface. Marine mammal predators counteract escapes by surrounding a school and trapping it at the surface as a "bait ball," exploitable by predator groups or individuals that dive through the ball. Actions are cooperative and coordinated; some individuals patrol the periphery, taking turns attacking the concentrated ball (Pierotti 1988b).

Marine mammals must search a large, complex habitat to locate concentrations of fish or squid, but they have limited vision. Pierotti's first research project suggested that sea lions, dolphins, and possibly humpbacks were using the presence of highly visible gulls, whose white colors are perceptible against all backgrounds at sea, to locate fish aggregations (Hoffman, Heinemann, and Wiens 1981; Porter and Sealy 1981, 1982; Pierotti 1988a). Gulls fly high when they leave island breeding colonies, forming widely spaced arrays with individual birds hundreds of meters apart in ragged "drive-lines." When a bird spots a school of fish, it spirals toward the surface, causing its white underside to flash, attracting the attention of other birds, which move toward the spiraling bird. As birds concentrate at specific locations, they attract the attention of other bird species, such as alcids, cormorants, and shearwaters, which normally fly close to the surface where they cannot see fish schools (Hoffman, Heinemann, and Wiens 1981; Pierotti 1988a). Gulls serve as catalysts of multispecies foraging groups (Hoffman, Heinemann, and Wiens 1981). Marine mammals use bird activity as cues to help locate fish schools, as do human fishermen. This is not simply exploitation by the mammals, however; the relationship is reciprocal because the mammals have the ability to concentrate bait balls at the surface, making the fish more available to the gulls, which cannot effectively plunge dive. Fish trying to escape from marine mammals often try to leap out of the water as an evasive tactic, allowing them to be readily captured by gulls (Pierotti 1988a, 1988b).

Some populations become dependent upon this cooperation. Pierotti observed gulls taking capelin (*Mallotus villosus*) that were driven to the surface attempting to escape feeding humpback whales off southeastern Newfoundland (Pierotti 1988a). This allowed herring gulls and black-legged kittiwakes to obtain capelin. A key observation that revealed the association between the seabirds and marine mammals was the timing of prey switching; breeding herring gulls did not start taking capelin until mid-June, even though capelin begin to arrive in Newfoundland's near-shore waters in May. Humpback whales breed around the Dominican Republic in winter (December through early March); in March they migrate back to their feeding grounds in the colder, more productive waters of the North Atlantic and arrive in Newfoundland in June (Pierotti 1988a). The dietary switch to capelin by gulls is therefore more strongly associated with the presence of whales than with the arrival of capelin themselves.

Human commercial fishermen also use avian foraging flocks to locate fish schools, often foraging in association with marine mammals. The notorious tuna/porpoise fishing controversy resulted from human exploitation of complex foraging relationships involving tuna, common dolphins (*Delphinus delphis*), and subtropical seabirds. The tuna industry learned to use bird presence to locate foraging aggregations. Off the west coast of North America, fishermen use the presence of gulls to locate schools of fish fed upon by sea lions. The human predators complain about the birds and mammals because they have the audacity to take fish while the fishermen are hauling their nets (Pierotti 1988b) and often become entangled in them, violating laws protecting marine mammals and seabirds.

Truly cooperative relationships have been observed between dolphins and Indigenous peoples in Brazil and West Africa. In Brazil this relationship has existed since 1847. Bottlenose dolphins, *Tursiops truncatus,* signal fishermen and drive fish toward shore into fishermen's nets (Pryor et al. 1990). This cultural tradition increases the foraging success of both species. Similar dynamics arose between orcas, *Orcinus orca,* and humans on the southeastern tip of Australia. Orcas drove migrating humpback whales close to shore, then signaled humans onshore, who rowed out to join them, and together the two species killed the whales. Humans took bones, blubber, and meat; the orcas seemed to want the fleshy tongues and lips of the whales, which are unused by humans. This relationship, which became known as "The Law of the Tongue," persisted for thousands of years. It started with Aboriginal peoples and was transferred to European whalers in the nineteenth century. The arrangement ended when some Europeans killed some young orcas, breaking established trust (Clode 2002).

The phenomenon of birds and mammals using each other's skill sets to exploit prey is not limited to marine ecosystems. Ratels, or honey badgers (*Mellivora capensis*), a large African mustelid, are led to beehives by honey guides (*Indicator indicator*), birds in the order Picidae. Honey guides also lead humans to beehives; after the humans remove parts of the hive to obtain honey, the birds feed upon larvae (Dean, Siegfried, and MacDonald 1990; Spottiswoode, Begg, and Begg 2016). It is possible that this behavioral interaction first appeared with the evolution of hominids, who knew how to use fire and smoke to counteract bee aggression.

COOPERATION AMONG RAVENS, WOLVES, AND HUMANS

Another clearly established bird/mammal relationship involves wolves and common ravens, *Corvus corax,* throughout much of North America (Heinrich 1999). Heinrich refers to ravens as "wolf-birds" (see also Munday 2013), a term which emerged from studies Heinrich did on ravens feeding on carrion in New England (Heinrich 1989). In Heinrich's study of populations in Maine, ravens were so shy near carcasses that he feared they "might be almost paralytically afraid of dangerous ground predators" (Heinrich 1999, 231), as suggested by a prominent behavioral ecologist (217). Heinrich traveled to Nova Scotia to observe a captive wolf pack and its interactions with ravens. He put down two piles of meat, one inside the wolf enclosure and one outside. Local ravens had a choice—feed with the wolves or away from them. In every case, ravens chose to feed with wolves, which suggested that ravens were not afraid of wolves but drawn to them.

To confirm his results, Heinrich looked at interactions between free-living wolves and ravens in Yellowstone National Park after the reintroduction of wolves there in 1994. In all cases he observed, carcasses attended by wolves also had ravens, magpies, and eagles feeding. Ravens never hesitated to feed from carcasses or meat piles attended by wolves (1999, 232–34). Ravens hung around wolves even when there was no food present, strongly suggesting that the ravens, rather than fearing predators, were shy and nervous around carcasses in the absence of wolves. As with the coyotes and badgers described above, wolves and ravens are "friends" in that they choose to associate. This does not mean, of course, that they are close emotionally, but there is trust and even mutual dependence. With wolves present, ravens and other birds are more confident and relaxed. The fear and shyness exhibited by the ravens in Maine resulted not from possible threats from terrestrial predators but from the absence of their mammalian companions. The behavioral ecologist who assumed that ravens were afraid was very much influenced by the "competition paradigm" described above. On the contrary, Cheyenne (Tsistsista) traditions of "calling" ravens, coyotes, and foxes to share in their kills (Schlesier 1987, 82; Hampton 1997; Fogg, Howe, and Pierotti 2015) can be recognized as part of a tradition that may go back millions of years to when raven and wolf ancestors shared similar interactions in the Pleistocene.

Are ravens simply scavengers assuming a "parasitic" role in the relationship? Durward Allen, a pioneer in ecological studies of wolves, observed ravens on Isle Royale following wolves (1979). Leading wolf behavioral ecologist Dave Mech describes ravens "chasing" wolves, flying above their heads, and observed a wolf and a raven playing "tag"—the raven pulled the wolf's tail, and the wolf then turned and stalked the raven (1970). According to wolf biologist Rolf Peterson: "There is more than playfulness between wolves and ravens. Ravens make their living by scavenging wolf-killed moose. . . . When wolves pause, the birds also stop, roosting in trees or swooping to the ice where they can harass the wolves at close range. Once disturbed, wolves resume travel [for hunting], which is what the ravens intended" (Peterson 1995).

These interactions start early in the life of a wolf; observations of ravens around wolf dens showed that whenever wolves got ready to hunt, they would have a group howl, which stimulated nearby ravens to become active and vocalize themselves. After pups emerge from the den at three weeks of age, ravens walk among them, and although they are large enough to prey on the pups, they only yank their tails, the same behavior they engage in with adults, suggesting that this tradition is established early in the pups' lives (Heinrich 1999).

Ravens locate carrion for wolves. Finding a carcass, they will start yelling, attracting other ravens, wolves, and coyotes. In addition, ravens guide wolves to moose or caribou vulnerable to predation. Ravens are unable to break the skin of frozen carcasses to reach the meat, depending upon the teeth and claws of wolves to provide them access. For the wolves, ravens serve as early-warning systems. According to Heinrich, "I can sneak up on a wolf, but never on a raven. They are incredibly alert" (1999, 138).

Having such relationships with wolves, ravens easily transfer this cooperative foraging to an equivalent alliance with humans. As Heinrich states, it makes no difference to a raven whether it is following a pack of wolves or a group of human hunters (1999, 243). Ravens monitor human hunting activities and assess human intentions: when Inuit hunters are going after caribou, ravens fly overhead, vocalizing and leading them to the prey. If Inuit are going out to gather berries, however, ravens seem indifferent and do not vocalize. Ravens develop similar relationships with Euro-American hunters, and they undoubtedly had such relationships with Plains tribes prior to the arrival of Europeans in North America. Ravens follow the best hunters, suggesting that they might have been especially attracted to mixed-species groups involving both humans and wolves.

GOING SOLO: A WILD DOG'S STORY

One key piece of evidence important to developing our story is examples of wild canids choosing to join social groups of different species. Such examples will demonstrate that the model we propose is part of the canid behavioral repertoire and can still be observed in the present. A compelling and insightful story fitting our requirements was observed in Botswana's Okavango Delta. An abundant population of African wild dogs, *Lycaon pictus,* decreased until only one small pack—two males and a female—remained. The two males disappeared, leaving the female alone to fend for herself. For six months, this female wandered the park hunting alone, avoiding other predators (National Geographic 2013).

Like wolves, *Lycaon* live in cooperative, highly social groups. After several months on her own, the female, now referred to as Solo, eventually gave up calling for her missing mates. Western scientists assumed she would wander away in search of a new pack to join, but Solo came up with an unanticipated, temporary solution by establishing social relationships with young spotted hyenas (*Crocuta crocuta*). Apparently feeling a need for some form of canid companionship, she also began associating with families of black-backed jackals (*Canis mesomelas*).

To many Western scientists, who regard these species as competitors ("enemies"), it is unlikely that a wild dog would form close bonds with hyenas and jackals, but this relationship worked, bringing benefits to all parties. Solo was a superb hunter who could chase down impala on her own. After forming cross-species bonds, Solo was accompanied on hunts by her mixed pack of hyenas and jackals. The impala did not recognize danger when the jackals accompanied Solo because jackals do not prey on impala, which made it easier for Solo to catch them, after which she shared the kill with her jackal "family." The other species obtained a reliable supply of food courtesy of her hunting skills, and they in turn warned Solo of approaching lions. So close did Solo become to her hyena friends that when she came into estrous she presented herself to them for mating (National Geographic 2013).

More interesting was her relationship with the jackals. Solo spent more time with the jackals, who initially were uneasy about having a larger predator species hanging around the den where they raised puppies. Initially, adult jackals tried to chase Solo off, biting and attacking her. Solo's solution was interesting: she inserted herself into the group by establishing bonds with the puppies. Eventually the parent jackal pair calmed down, seeing that no harm came to the pups. Solo frequently regurgitated food to the jackal puppies just as she would with her own pups, and the jackals accepted her. Solo spent days resting close with the jackals, and the pups grew extremely close to her.

During the dry season, impala leave the edge of the swamp, which forced Solo to leave her jackal family in search of food. Encountering a new jackal family with tiny pups, she effectively kidnapped them, preventing the parents from getting near them. The adult jackals were upset at first but relaxed as they saw no harm coming to their pups, who bonded with Solo as she fed, groomed, and protected them, even chasing away her hyena friends when they wandered close. Solo was essentially imposing pack structure on a species that does not form packs or engage in group hunting in order to meet her own social needs. When a large *Lycaon* pack came by, Solo hid and stayed with the jackal family, continuing to function as a provider for her adopted family of another species.

CONCLUSIONS: HUMANS AND WOLVES

If we substitute *Canis lupus* for *Lycaon pictus* and *Homo sapiens* for the jackals, we have a scenario that might lead to the establishment of social bonds between wolves and humans. Wolves often become solitary when they either leave or are driven from their pack. They always seek companionship because lone wolves do not usually survive long on their own. The scenario posited in our introduction of a solitary pregnant female is quite similar in its basic dynamics to the behavior of Solo, a real-life parallel. A female canid of reproductive age, forced into a solitary situation, joined social groups of other carnivore species somewhat similar to her in their ecology and behavior. Female wolves who become pregnant when alpha females have puppies will almost certainly have their pups killed by the dominant female (McLeod 1990), so she must disperse, alone and pregnant. Under the circumstances, it seems completely logical for such a female to join a group of social predators, just as Solo did.

When early humans first encountered wolves after leaving Africa and moving into Eurasia around 40,000 years ago, the two species appear to have established a partnership that allowed *Homo sapiens* to eventually dominate the entire world (Allman 1999). DNA evidence from both modern dogs and humans suggests that these events occurred at roughly the same time (Germonpré et al. 2009; Ovodov et al. 2011; Druzhkova et al. 2013). This early partnership could have enabled *Homo sapiens* to displace the other competing hominids—the Neanderthals of Europe and Denisovans and *Homo erectus* of Asia—as they proliferated throughout the habitable areas of the world (Shipman 2015). Human alliance with wolves/dogs enabled humans to expand into these inhospitable areas and ultimately invade the New World (Allman 1999). Humans both profited and learned from wolves, long before any bonding or domestication took place.

Wolves probably taught cooperative hunting to humans, giving them advantages over other species of *Homo* (Schlesier 1987; Fogg, Howe, and Pierotti 2015; Shipman 2015). Wolves and dogs are unique among animals in the nature of their cooperation; they don't compete but form friendships (Lienhard 1998–99). An ape may bond with its offspring but, as illustrated by Marshall Thomas (1993, 2006), dogs form alliances beyond kin. Wolves are inclined to cooperate with other species, for example, ravens and even bears (Edwards 2013). Human and wolf joined forces in the hunt, probably changing human behavior in ways that may have made us the beings we are today (Schleidt and Shalter 2003; Shipman 2015).

Cooperative foraging means that two species have become important aspects of each other's ecological niches. When populations of organisms construct ecological legacies that are continued over many hundreds of generations, they modify selection pressures acting upon subsequent generations. These altered selection pressures favor the driving trait, causing these traits to spread through future generations. When niche construction processes have evolutionary consequences that perpetuate these ecological legacies, this can be described as *ecological inheritance* (Shavit and Griesemer 2011).

Such a dynamic was involved in the interaction between humans and wolves. Each species affected the other, allowing both to invade ecological niches that would not have been possible without their heterospecific companions, such as killing mammoths (Shipman 2014, 2015). After leaving the tropics, humans were able to persist in areas where climates were less predictable and more extreme because they had a companion species well adapted to such conditions. In return, wolves adjusted to life alongside humans, which allowed them to outcompete all other carnivores, including bears and big cats, and to take larger prey in greater numbers than before. As their feeding environments became more consistent, wolves were able to alter their reproductive tactics and behavior, leading eventually to their special position as the number one companion of humans all over the globe (Morey 2010).

That this relationship is cooperative at its root does not mean that conflicts could not arise. At times violent clashes might result, which we see in the way Inuit and Inupiat people sometimes treat their sled dogs (Shipman 2015). Individuals, even within families, still act selfishly at times, the source of many conflicts. Cooperation and conflict should be seen as two sides of the same coin, either of which can emerge when organisms have social relationships and live in close proximity. Contemporary examples can be seen every week on Cesar Millan's TV show *Dog Whisperer,* whose fundamental premise relies upon conflicts between contemporary *Homo sapiens* and domestic versions of *Canis lupus.* It is even more likely that conflicts arose between nondomestic *Canis lupus* and the aggressive, powerful hunters and gatherers who were early *Homo sapiens.*

CHAPTER THREE

Homo canis

WHY HUMANS ARE DIFFERENT THAN ALL OTHER PRIMATES

IN THIS CHAPTER, WE EXPLORE the history of *Homo sapiens* and the forces that caused this species to evolve into an organism more imaginative and flexible in its behavior and ecological responsiveness than any predecessor. As modern humans expanded out of Africa into Eurasia, they encountered new environmental conditions and unfamiliar species, including *Homo neanderthalensis,* and possibly even *Homo erectus* in southern Asia (Harris 2015; Shipman 2015). Over time, *sapiens* displaced these related forms and established complex relationships with other species, the first of which was almost certainly *Canis lupus.*

This flexibility does not necessarily mean *Homo sapiens* was superior to other life-forms; in fact, they narrowly dodged extinction on several occasions (Harris 2015). Western cultural traditions have long assumed human superiority combined with separation from other life-forms (Pierotti 2011a, 2011b). There was a nineteenth-century movement in England to place humans in their own order, Bimania, because the "particular characteristics of man appear . . . so strong," particularly the erect posture and distinctive hind feet of humans in contrast with the "four-handed" apes and monkeys (Ritvo 2010, 181). Lest we think that modern science has advanced, there is reluctance among many contemporary systematists to locate humans and close extinct relatives within the family Pongidae. Seeking differences, we often lose sight of similarities between ourselves and other life-forms that establish that we are of the Earth, related to all others (Pierotti 2011a, 2011b; Pierotti and Wildcat 2000).

We share conserved processes that operate at the cellular level (Kirschner and Gerhart 2005), including the use of DNA and RNA in replication, the ability to derive energy from breaking down sugars that is then captured and stored in the form of phosphate bonds. All living entities use the same basic molecular mechanisms and cellular processes that appeared over 3 billion years ago in the earliest forms of life. These have existed almost unchanged for billions of years, and DNA sequences that exist in one-celled Archaea today are very similar to those found in ourselves, even though we and other eukaryotes have considerably more genetic material.

Among the numerous forms of multicellular eukaryotic life are animals, of which one category is the vertebrates (animals with backbones). Within vertebrates are mammals, including organisms as diverse as whales, bats, cats, mice, deer, cattle, and primates. Humans are in the order Primates, which includes lemurs, tarsiers, marmosets, monkeys, and apes. Genetically these are our closest relatives, especially great apes: chimpanzees, bonobos, gorillas, and orangutans. The ape family Pongidae should include humans, who have been characterized as "the third chimpanzee" (Diamond 1992). Despite scientific "objectivity" and the drive to "see their work in isolation—unconstrained by their own context . . . despite their careful definitions and their forced assertions, scholars are inevitably influenced at least as much by the common usage of the terms that they deploy, as they are by their more rarefied and specialized senses" (Ritvo 2010, 4). Thus the genus *Homo* and its extinct relatives in the genus *Australopithecus* are segregated into the family Hominidae, although we *Homo sapiens* share around 98.8 percent of our genome with chimpanzees, *Pan troglodytes* (Harris 2015). This 1.2 percent chimp-human distinction involves only a count of substitutions in the base-pair building blocks; comparison of the entire genome, however, indicates that segments of DNA have also been deleted, duplicated over and over, or inserted from one part of the genome into another, creating an additional 4–5 percent distinction between human and chimpanzee genomes (Harris 2015). Other species of apes are now included within the Hominidae, so at least this approach considers humans and our ape cousins to be members of the same family.

Despite genetic similarities, our primary interest in this book is in the social systems of primates and how these relate to environmental factors such as food supply. Primates show a wide diversity of social systems (Kappeler and van Schaik 2002), including spacing of individuals, group size, and mating pattern as well as regarding patterns and quality of social relationships. Diversity is evident not only among species but can also be found within both species and populations (Richard 1978; Goldizen 1987).

Variety among social systems within species or populations suggests that these systems are not genetically driven but represent *emergent properties* of individual behavioral interactions and strategies (Hinde 1976), an important concept in understanding behavioral evolution. It is often assumed that variation in behavior among individual animals is controlled by genetic variation (E. O. Wilson 1975, 1978, 1994; Dawkins 2006). The concept of emergence is more sophisticated, recognizing that behavior, especially at the individual level, results from response to specific environmental conditions faced by individuals (O'Connor and Wong 2012). The term *emergent* means newly appearing or "happening right now" because examining all the parts of a biological system without considering the whole means you will not see what the system is capable of producing (Lewontin 2001).

Social dynamics are shaped by ecological factors—for example, distribution of risks and resources—and interactions among such factors (Kappeler and van Schaik 2002). Most primate groups have multiple adults of both sexes; however, in other mammals permanently bisexual groups are rare (canids are another exception to this general pattern). There are four basic mating systems: polyandry (one female, multiple males), polygyny (one male, multiple females), polygynandry (multiple males, multiple females), and monogamy (one male, one female). All are found in primates (Crook and Gartlan 1966; Eisenberg, Muckenhirn, and Rudran 1972; T. H. Clutton-Brock and Harvey 1977). Canids are one of the few mammalian groups where this is also true; however, canids show complex relationships among body size, group size, and mating system. Smaller, less social species, for example, foxes, tend toward polygyny within monogamy, female helpers, and male dispersal, whereas larger species such as wolves show polyandry within monogamy, with male helpers and female dispersal (Moehlman 1989).

In primates, variation in number of adult males is an obvious feature, with a general distinction between single- and multimale groups. There are no canid social groups with multiple adult males. This dichotomy is not a species-specific trait but a flexible response to variation among groups in food supply or male and female survival rates (Kappeler and van Schaik 2002). Reproductive synchrony among females is one important determinant of the outcome of the dichotomy (Altmann 1990).

Multimale groups typically are found with large numbers of females (Mitani, Gros-Louis, and Manson 1996). Group size, also an emergent property, varies widely (Kappeler and van Schaik 2002). Primate groups vary in size by orders of magnitude (Dunbar 1988; Kappeler and Heymann 1996), with four main correlates impacting variation in group size. First, widely distributed feeding areas generate increased foraging and travel costs, setting upper limits on group size because of increased intragroup feeding competition (van Schaik 1983; Janson and Goldsmith 1995). In contrast, larger group sizes and multi-male and -female groups are favored by reduced predation risk, combined with intense intergroup feeding competition (Wrangham 1980; van Schaik 1983; van Schaik and van Hooff 1983). Risk of infanticide by controlling males favors reduced group size (Crockett and Janson 2000; Steenbeek and van Schaik 2001). Size of the cerebral neocortex limits group size because it determines ability to process complex information about social relationships (Dunbar 1992, 1995, 1998).

The least gregarious primates show solitary, dispersed social systems, in which an adult male's territory overlaps those of one or more adult females, but each individual forages alone and maintains social contact mainly through vocal and/or olfactory communication. Most are nocturnal, foraging at night and sleeping in trees during the day. Mating systems in these primates are usually polygynous among prosimians and, oddly enough, in orangutans (*Pongo pygmaeus*), the only ape with a solitary social system (Dunbar 1988; Swedell 2012).

Some primates show pair-bond-based social systems, in which an adult male and female form a small social group with their offspring, defending a territory against other pairs. This category includes several South American monkeys, and gibbons and siamangs from the apes. Mating systems in these groups are monogamous, although extra-pair copulations have been observed. With the exception of gibbons, males in pair-bond-based groups participate in offspring care, which is unusual for male mammals (Dunbar 1988; Swedell 2012).

A common mammalian social system is single male groups, found in colobine monkeys, guenons, patas monkeys, howler monkeys, geladas, Hamadryas baboons, and sometimes in gorillas. A single resident adult male defends a group of philopatric, related females against other males and, while his tenure lasts, enjoys exclusive mating access typically involving polygynous mating systems, accompanied by sexual dimorphism. Sometimes called "harems," these groups are always at risk of being taken over by nonresident males, who form all-male groups while awaiting their chance to become resident males (Dunbar 1988).

Multimale and multifemale groups found among baboons and macaques, where multiple individuals of each sex form large social groups, represent the largest, most complex primate groups. These groups are complex socially, showing differentiated social and kin relationships among group members. Females remain with the group throughout their lives, whereas males disperse upon attaining maturity (Dunbar 1988). A variation on such groups are fission-fusion communities, which are less cohesive because such groups occupy large home ranges in which temporary foraging parties cleave and coalesce over time according to changes in resource availability and female reproductive condition. Fission-fusion communities are characterized by female dispersal and male philopatry (Swedell 2012).

The most complex type of social system found in primates, and in mammals overall, is the multilevel society (hierarchical or modular societies) characterizing Hamadryas baboons, geladas, snub-nosed monkeys, and a few other mammals, including elephants. In such systems, there are at least three levels of social structure: one-male units (OMU), bands, and troops or herds. OMUs are the reproductive unit, consisting of one "leader" male, sometimes a follower male, and several females. Bands are ecological units that forage and sleep together, and troops are large, temporary aggregations at sleeping sites or foraging areas. In Hamadryas baboons, a fourth layer exists between OMU and band: the clan, consisting of OMUs and bachelor males linked by social bonds. In geladas (and elephants), bachelor males form all-male groups. Reproduction is usually polygynous (Dunbar 1988).

Diversity in social and mating systems is apparent among our closest living relatives, the apes. Social systems range in complexity from large, open-unit groups of chimpanzees and bonobos to single-male gorilla groups to mated gibbon pairs to solitary orangutans (Mitani 1990). If humans are the third chimpanzee (Diamond 1992), this suggests that a common ancestral state for Miocene hominoids was living in chimp- or bonobo-like multimale/multifemale groups with a tendency to undergo fission and fusion in response to ecological and/or social variables (Malone, Fuentes, and White 2012).

Among primates, humans are unusual in two ways. First they are used to extending social interactions beyond kin and the immediate social group. Second, they show extensive intraspecific flexibility in social organization, which means they can rapidly adjust to new ecological situations. An essential feature of hominoid evolution is the shift from limited plasticity in most generalized social apes to expanded behavioral plasticity. In addition, the potential for innovation, spread, and inheritance of behavioral patterns and new social traditions through cultural transmission is much higher in the hominoids, especially the great apes, than in other anthropoid primates (monkeys and baboons), which might provide a basis for substantial expansion of social complexity and behavioral plasticity in the genus *Homo* (Malone, Fuentes, and White 2012).

Sociality in apes and humans is related to sexual dimorphism between males and females, with polygynous species showing males that are much larger than females, as is true of most mammals. To assess dimorphism in extinct forms of hominids, the usual practice is to assess relationships between body and skeletal size in specimens of known body weight. Findings are then applied to hominid fossils using a variety of statistical methods, knowledge of the associated partial skeletons of early hominids, formulae derived from modern humans, and common sense (McHenry 1992, 1996). In *Australopithecus afarensis*, females are 64 percent the mass of males; in *A. africanus*, 73 percent; in *A. robustus*, 80 percent; in *A. boisei*, 69 percent. In the earliest *Homo* species, *H. habilis*, females are 62 percent the mass of males. Overall body sizes remained small relative to modern humans, but there was marked sexual dimorphism, implying polygynous mating systems. Between 2.0 and 1.7 million years ago (MYA), there was a rapid increase to essentially modern body size with the appearance of *Homo erectus*, which also showed sexual dimorphism greater than seen in modern humans but below that seen in modern gorillas, chimps, and orangutans, all of which are clearly polygynous. This implies social organizations in Australopithecines characterized by kin-related, multimale groups with females who were not kin-related—that is, female dispersal. In contrast, modern humans are very different than these polygynous species, with reduced sexual dimorphism and a tendency for males and females to form pair-bonds.

Two divergent trends exist in relative sizes of cheek teeth, which is also an indicator of overall dimorphism. There is initially a steady increase in tooth size in Australopithecines from *A. afarensis* to *A. africanus* to the "robust" australopithecines, followed by a decrease beginning with the appearance of the earliest *Homo: H. habilis* to *H. erectus* to *H. sapiens* (McHenry 1992, 1996). This pattern reflects profound changes leading to the increased intelligence and social flexibility of modern humans. Early *Homo, habilis* and *rudolfensis,* retained primitive features shared in common with australopithecines; however, both shared key unique features found in later *Homo,* for example, increased size of the brain and decrease in masticatory apparatus (jaw muscles and bones) relative to body mass (McHenry and Coffing 2000).

Analysis of human gene sequences reveals a mutation in the class of genes that code for myosin heavy chain (MYH), which are associated with changes in jaw structure and tooth size (Stedman et al. 2004; Harris 2015). MYH are a critical protein component of the sarcomeres, the "engine room" of skeletal muscle, from which contractile force is derived. Different MYH molecules are specialized for different muscle-contraction rates specific for different muscles. Inactivation of individual MYH genes results in dramatic reduction in the size of the muscles in which they are active. The particular gene MYH16 is specifically expressed in primate jaw muscles, including in humans. The mutation of interest prevents accumulation of MYH16 protein. Stedman et al. 2004 found that all nonhuman primates for which genome sequences could be obtained have an intact copy of the gene and show high levels of MYH16 in their jaw muscles. Jaw muscles with high levels of MYH16 are so powerful they require ridges atop the skull called sagittal crests, which serve as anchor points for large jaw muscles.

Decrease in jaw-muscle size produced by inactivation of MYH16 removed major barriers to the remodeling and expansion of the hominid cranium, consequently allowing an increase in brain size. This mutation seems to have appeared about 2.5 million years ago, the period just before evolution of the distinctive cranial form found in the genus *Homo*. In great apes, the massive jaw muscles are attached to the sagittal crest, which is so well developed that it constrains growth of the cranium, thus limiting brain growth. Inactivation of MYH16 removes the requirement for a sagittal crest to serve as an attachment site and anchor for the jaw musculature because jaw muscles are smaller and weaker. Animal models of jaw-muscle transposition or removal show that changing muscle anatomy radically alters growth patterns (Harris 2015), including reduction in stress across the bones of the braincase, allowing it to expand. With no requirement for a sagittal crest, the cranium can expand and become rounded rather than peaked, allowing room for the brain to expand, especially cerebral hemispheres. What is significant about this finding is that there appear to be no concurrent specific mutations favoring enlargement of the cranium, although changes in genes favoring increased cranial capacity could well have followed jaw-muscle mutation release of physical constraints on cranial structure (Kirschner and Gerhart 2005).

The implications of this finding provide an understanding of how evolution functions. Mutation in jaw musculature was almost certainly not favored initially by natural selection; in fact, it was probably strongly selected against. Individuals with weak jaw muscles would be severely constrained in diet, unable to handle tough foods readily handled by their ancestors who lacked the mutation in muscle cell structure. Thus, one crucial change that led to evolution of the genus *Homo* was almost certainly a major selective disadvantage upon its first appearance. Only a few individuals carrying this mutation might have survived, which could explain a population bottleneck found to exist in early human evolution (Harris 2015).

Around the time when this mutation arose, stone tools first appeared, perhaps as a result of new mental states emerging along with the expanding unconstrained cerebrum. Associated with these cranial changes, body size increased, sexual dimorphism decreased, and limb proportions changed. Weaker jaw muscles led to reduction in cheek teeth size, and crania began to show some unique features—for example, the domelike shape that we associate with later *Homo* (McHenry and Coffing 2000). This mutation appears associated with major changes in both human physiognomy and behavior. The time period around 2.5 MYA cycled rapidly between extremely wet and extremely dry periods (Gibbons 2013). Such conditions would have favored an organism that was flexible in behavior and ecology, and probably favored large-brained generalists over more specialized, less flexible ancestral forms. Similar mutations in muscular structure may have occurred many times in the past; however, individuals carrying such mutations probably went extinct. It is possible that 2.5 million years ago, conditions were right for individuals carrying this mutation to persist and subsequently to thrive because the initial negative selection pressure would have quickly moderated as the benefits of the released constraints on brain size and behavioral flexibility and innovation in a rapidly changing environment emerged. This represents an example of the sort of phenomenon associated with punctuated evolutionary events: small genetic change leading to major changes in ontogeny, producing a new form (Eldredge and Gould 1972).

Significant internal brain reorganization was associated with changes in cranial structure between *Australopithecus* and *Homo*, although the fossil record preserves only the exterior shape (McHenry and Coffing 2000). These findings can be summarized in the following scenario. Before about 2.5 MYA, stone tools were absent at sites containing hominid fossils, brain sizes in Australopithecines were comparable to those of chimpanzees, cheek teeth and supporting masticatory structures were large and powerful, and primitive traits were retained in all parts of the body, including the skull, overall body size was small, there was strong sexual dimorphism in body size, and hind limbs were small relative to forelimbs. By 1.8 MYA, *H. ergaster*, the first member of the genus *Homo*, appeared, distinguished by more humanlike body and behavior (Harris 2015). At this point, we have the appearance of a new genus of apelike primate, *Homo*. This genus would eventually evolve into contemporary *Homo sapiens*, which appeared in Africa about 200,000 YBP (0.2 MYA) and spread into other parts of the world over the last 100,000 years (Harris 2015). As a result of this dispersal, *Homo sapiens* encountered recent ancestors of modern *Canis lupus*, setting in motion events that led to contemporary human and canid companionship (Shipman 2015).

We can only surmise what social structures other hominids had based on our knowledge of contemporary primates and evidence of sexual dimorphism from the fossil record (Harris 2015; Shipman 2015). Given social structures found in other apes, it is unlikely that early *Homo* lived in single-family groups. Humans' closest living relatives are all sexually dimorphic species: polygynous gorillas, chimpanzees, and orangutans (J. Clutton-Brock et al. 1977; T. H. Clutton-Brock and Harvey 1977; Dunbar 1988, 2000). In polygynous species reproductive success varies more among males than among females (Trivers 1972). Higher degrees of intermale competition relative to interfemale competition select for traits in males likely to lead to success in combat—for example, large body and large canines—which, in turn, causes the size and shape of male and female bodies to diverge. Other traits found to covary with sexual dimorphism include differences in parental investment, patterns of resource acquisition and utilization, and division of labor (Trivers 1972; Pierotti 1981; Annett, Pierotti, and Baylis 1999).

Neanderthals were less sexually dimorphic than *Homo sapiens* (Quinney and Collard 1995; Harris 2015; Shipman 2015). Modern humans are less dimorphic than African *Homo sapiens'* ancestors, suggesting that modern humans evolved to show less dimorphism and greater monogamy than our very recent ancestors, even if we retain the capacity for showing occasional polygyny or polyandry depending upon environmental conditions and local sex ratios. We have little under-standing of the social structures of Neanderthals or Denisovans, our closest relatives within the genus *Homo*. In fact, we don't really know that much about the social structures of our *sapiens* ancestors. What we do have is information on the social structure of contemporary *Homo sapiens* around the globe. The majority of contemporary humans live in social groupings that approximate the structure of a wolf pack or family group. As mentioned above, canids generally show monogamy as their primary mating system, with occasional instances of polygyny or polyandry (Moehlman 1989) which, at the least, suggests convergence at the behav-ioral level between modern humans and wolves.

A pack, or family group, of wolves typically consists of the same cast of characters found in an extended human family. There is an alpha female, along with her mate and their offspring of various ages (Moehlman 1989; Pierotti 2011a; Spotte 2012). Offspring, especially females, stick with their maternal pack after reaching adulthood. Older offspring help rear later offspring, increasing the growth of the pack. Modern humans are the only great apes that regularly practice monogamy which, because mating systems are emergent properties shaped by environmental influ-ences, could well have been a trait picked up from wolves, or at least a trait that both species share because of convergent ecological niches. Both species share similar ecological roles as efficient group-hunting predators taking large prey; both benefit from cooperative food sharing. Humans adopting lupine social strategies could explain why wolves are easily assimilated into human social groups, and the reciprocal situation has also been reported in many cultural traditions, for example, Romulus and Remus, the *Jungle Book,* and many stories from American Indigenous peoples (Bettelheim 1959; Itard 1962; Singh and Zingg 1966; Lane 1976).

Another factor supporting the adaptiveness of the relationship between humans and wolves can be seen in the difficulties that arise when chimpanzees, *Pan troglodytes,* are adopted or cross-fostered by humans (Fouts and Mills 1997). From the 1950s through the 1970s, adoptions of young or infant chimps by humans were fairly common in the United States. Such adoptions worked until the young chimps were about five years of age, at which point many human parents decided they could no longer care for their foster "offspring," and some infant chimps died of stress when separated from their foster mothers (Fouts and Mills 1997). In the wild, young chimps maintain a strong association with their mothers until they reach adulthood, with some individuals maintaining the bond well past that point. In contrast, young wolves typically disperse to form their own groups and can integrate themselves easily into human groups because they do not form bonds only with a single individual. If they wish they can spend their entire lives within a human family group, as is seen with many canids, both today and throughout human history.

Existence of monogamous pair-bonds in humans is significant; most primate social groups lack pair-bonds. Among apes, only gibbons show pair-bonds, which are completely different than those seen in humans; males defend territories on which females and offspring live but have little interaction with the offspring and provide little care (Mitani 1990; Kappeler and van Schaik 2002). Our modern human ancestors were different from contemporary humans in many ways. They had little or no fear of "nature" and their environmental surroundings. This attitude is seen in very few modern humans of European ancestry, who regard nature as a hostile and dangerous place, but has been observed in Indigenous peoples around the world (Sale 1991; Pierotti 2011a).

The standard account of the origin of modern humans is that they originated somewhere in Africa, about 200,000 YBP, with evidence for single origin based on the idea of a "mitochondrial Eve," or most recent common ancestor, that is, a single female who was the founder of the mitochondrial lineage found in all modern humans (Cann, Stoneking, and Wilson 1987). We can see the influence of Christian thought on Western science, with the use of the Garden of Eden myth to generate a metaphor for this "scientific" discovery. Recent study of evolutionary history of nuclear genes reveals a much more complex history behind the origin of *Homo sapiens* than a simple single origin (Harris 2015). Ironically, the history of the evolution of *Homo sapiens* bears considerable resemblance to the process by which wolves became domestic dogs— there were multiple populations that subsequently interbred with one another in Africa, producing a variety of phenotypes (Garrigan and Hammer 2006; Hammer et al. 2011).

Modern humans moving out of Africa experienced a second, more serious population bottleneck around 50,000 years ago, presumably because they had to adapt quickly to new, previously unexperienced environmental conditions, including new diseases (Harris 2015). One adaptive response was to interbreed with other hominids encountered in Eurasia, both Neanderthals and Denisovans, who provided some genetic resistance to new diseases (Krings et al. 1997; Harris 2015). Genomic analysis estimates that interbreeding between Neanderthals and modern humans took place around 50,000 years ago (Sankararaman et al. 2012). Denisovan DNA has also been found in modern human populations living in Australia, Papua New Guinea, the Philippines, and Indonesia (Harris 2015). This is hardly surprising because hybridization is a common response of populations experiencing new or changing environmental conditions, especially in lineages that show monogamous breeding systems (Grant and Grant 1992, 1997a, 1997b; Pierotti and Annett 1993; Good et al. 2000).

Another adaptive response by modern humans moving out of Africa into the colder and more variable Eurasian environments was the establishment of cooperative hunting, and probably living, relationships with wolves around 40,000 YBP (Schlesier 1987; Schleidt and Shalter 2003; Shipman 2014, 2015). Early humans may have recognized the need to learn the worldview of another species as well as to accept their understanding of how to survive under new conditions, realizing that the other species' knowledge was more reliable, especially when hunting unfamiliar prey. Most contemporary humans cannot conceive of such a relationship, although acceptance of nonhuman perspectives is clearly present in pre-contact Native Americans (Pierotti 2011a, 2011b; Fogg, Howe, and Pierotti 2015), Australian aboriginal peoples (Rose 2000, 2011), and even people working successfully with border collies (McCaig 1991).

Such attitudes seem more common in parts of Europe where people live easily with dogs. This perspective is much less common in America, where people are often uneasy about dogs (McCaig 1991). Marshall Thomas goes further: "Wolves and their descendants have reached a climax of domestication although . . . we in the United States must travel elsewhere to see it. Europeans are vastly more civilized than Americans when it comes to dogs—too many Americans are veritable dog fascists whose need to control is so overwhelming that they would have all dogs caged if they could" (2000, 128–29). We understand Marshall Thomas's point from our involvement with larger breeds of dog and alleged "wolf-dogs." Anyone doubting this interpretation need only watch a few episodes of Cesar Millan's television show *Dog Whisperer* to see how constantly anxious Americans are about, and often in open conflict with, their canid pets, which only succeeds in causing stress to the poor dogs.

There are continual efforts to find the difference between humans and all other animals (Pierotti 2011a). Differences exist, but these are almost certainly quantitative rather than qualitative—that is, extreme expressions of traits found in a variety of other species. An example can be seen in a recent effort to compare "societies of wolves and free-ranging dogs" (Spotte 2012). One key argument underpinning Spotte's lengthy text is that only humans have a theory of mind—that is, they can attribute motivations to other animals with whom they interact (see Pennisi 2006). Humans are the model for this concept because we know that humans can recognize the difference between themselves and others. Spotte argues, "Unlike us, animals are unable to recognize the mental states of others. Consequently, senders in the animal world *do not intentionally signal to inform others,* and receivers interpret a sender's signal *without recognizing it as reflecting what the sender knows*" (2012, 74).

Given the evolutionary relationship we have been discussing, this seems an odd assertion to make concerning canids, who have had to show great sensitivity and understanding of human social communication and mental states to allow this cross-species dynamic to work. It specifically involved communication and intentional signals not only within but between different species as well as recognition and interpretation of signals sent by others. Anyone who has owned a border collie, among many other breeds, knows that *Canis lupus* reveal very complex mental states. Theories of mind are often identified, or at least surmised, when individuals of a species display the ability to recognize themselves in a mirror, which has been found in elephants, dolphins, apes, and several birds, including ravens, magpies, and parrots (Pennisi 2006).

Marshall Thomas herself reveals how a contemporary well-educated and sophisticated New Englander can come to understand relationships between different species, but she is an unusual person, having spent several years in her late teens and early twenties living with the Ju'wasi people in the Kalahari Desert, acquiring a good understanding of how Indigenous peoples see the world (Marshall Thomas 1994, 2000, 2006). One strength of Marshall Thomas's work is her ability to recognize how nonhumans perceive the world and consider how they might think about it. This provides an advantage over scholars who read every paper ever published in the peer-reviewed literature but have little or no feeling for what nonhuman animals are truly like (e.g., Spotte 2012). Over her career, Marshall Thomas has observed behavior in lions, elephants, deer, and "domestic" cats and dogs. Mixed in among these varied experiences was a period spent on Baffin Island watching—or perhaps it is better expressed living with—a pack of wolves (1993, 2013). Her experience seems like the sort that modern humans might have had when they first encountered wolves after emigrating from Africa.

One reason for this assumption is that the wolves on Baffin Island were not used to humans and had never been hunted. Marshall Thomas's observations took place during July and August in the high Arctic, so there was no night. In her first encounter, "two white wolves came trotting around a hill and saw us. One was an adult, who increased her pace and disappeared over a ridge. . . . The other was juvenile, and he kept right on trotting as before, doing exactly what the Ju'wasi had told us to do in the presence of a predator—don't run, just move away in a moderate manner at an oblique angle" (2013, 219). This advice is useful for anyone confronted with a potential predator or a fellow carnivore. You want to appear calm and confident, so you move away without running and without turning your back, which shows you are respectful but not afraid, whereas turning around and running suggests that you might well be prey, making you vulnerable.

These wolves had four dens, which may well have been occupied for thousands of years and hundreds of wolf generations, since the end of glaciation allowed the wolves to colonize the island. Marshall Thomas spent several weeks alone observing this pack and never felt threatened, even though the pack members did seem curious about her, investigating the cave where she camped and any objects she left lying about. If we imagine a family group of *Homo sapiens* encountering wolves for the first time, we can picture a similar dynamic—curiosity on the part of both species, mixed with respect and caution based on the recognition that both species were group-living, and hence formidable, predators. Marshall Thomas indicates that there seems to be a sort of automatic protocol shown by predators when encountering one another unexpectedly that serves to minimize potential conflicts.

Marshall Thomas learned how to behave around predators through experience with lions, the largest social predator on land (1994, 2006). Humans leaving Africa, dispersing to other landmasses, probably also had learned in this manner. Lions have never been domesticated, however, and are not a species that hunts cooperatively with other species, although it seems the Ju'wasi had a sort of truce with local lions, such that the two species avoided one another and worked to minimize competition (Marshall Thomas 1994).

The first wolves to encounter modern humans may have chosen to interact in a very different manner. These wolves would have been familiar with other members of the genus *Homo,* such as Neanderthals and Denisovans, or even *Homo erectus,* who had been living alongside them for many thousands of years, seemingly without forming any sort of social bonds and long-term interactions, or at least bonds of the sort that led to obvious phenotypic changes in wolves. Some argue that one reason *Homo sapiens* replaced larger, more physically powerful *H. neanderthalensis* throughout Europe and Asia was that *sapiens* had wolves as partners and Neanderthals did not (Coren 2006; Shipman 2015).

Other scholars think that these early *Homo* may well have had cooperative relationships with wolves, but the interaction never generated any physical changes that could be recognized in the archaeological record until the last 15,000 years or so (Schleidt and Shalter 2003; Derr 2011). "Dogs/Wolves changed behaviorally thousands of years before they produced a distinctive phenotype. Whatever form these initial relationships assumed, it is very unlikely that they involved 'selective breeding' to produce a desired phenotype" (Coppinger and Coppinger 2001, 50). Wolves and humans coevolved, producing primarily behavioral and ecological (emergent) changes in each other's ecology and behavior. Each species changed the basic nature of the other without necessarily changing its appearance, beyond a possible reduction in human sexual dimorphism. It is interesting that the posited extinction time for both Neanderthals and Denisovans (about whom much less is known) appears to be around 30,000 YBP, the same time that possible domestic wolves (dogs) appear in the archaeological record (e.g., Germonpré et al. 2009, 2012, Ovodov et al. 2011). Prey taken in northern Asia by Paleolithic humans were primarily species also taken by wolves, wild ass, wild sheep, bison, and reindeer, with little evidence of mammoth or rhino hunting (but see Shipman 2014, 2015). Archaeological records suggest that in Siberia these people employed short-term small camps, possibly involving men and wolves loosely linked to larger, more stable base camps dominated by women (Goebel 2004; see chapter 5).

The model proposed by Schleidt and Shalter 2003 takes a new spin on the Hobbesian theme of *Homo homini lupus:* "Man to man is [both a] kind of god [and] an errant wolfe" (Hobbes 1985, quoted in Schleidt and Shalter 2003). They argue that, given the strong social bonds between humans and canids, "Man to man is—or at least should be—a kind wolf," and that recent changes in public perception of wolves allow us an opportunity: "instead of perpetuating our traditional attitude that 'domestic animals' are intentional creations of human ingenuity," we might consider the possibility that "initial contacts between wolves and humans were truly mutual, and that various subsequent changes in both wolves and humans must be considered as a process of co-evolution" (Schleidt and Shalter 2003, 58), a proposition that refutes the argument proposed by Francis (2015). More significantly, they argue that wolves may have had as great an impact on human evolution as humans have had on wolves in accommodating the change of some wolves into dogs.

Their arguments are based upon estimates of the length of time during which humans and wolves may possibly have been interacting.

One area of difference between our views and those of Schleidt and Shalter (2003) is that they contend that the relationship between the genus *Homo* and wolves could actually have begun with *Homo erectus* as far back as 400,000 YBP. Although interactions between early *Homo* and wolves may have been benign, little archaeological evidence exists concerning such social relationships. There are, however, indications that human social and even ethical systems—which many admire and hold, at least in theory, to be the highest achievement of humanity—were invented first by early canids. Cooperative, unselfish behavior with minimal conflict is practiced to this very day by some of their descendants and honed to perfection by members of the pack-hunting canid species, notably the gray wolf, African wild dogs (*Lycaon pictus*), and the Asiatic wild dog (*Cuon alpinus*) (G. K. Smith 1978).

Support for this argument exists in the observation that *Homo sapiens* and *Canis lupus* are the two most widely spread terrestrial mammals, capable of surviving in practically any environment: from polar tundra, taiga, or deciduous forest to grasslands and even deserts. The ability of these species to survive under conditions of extreme seasonality allows them to deal better with temporal fluctuations in environmental conditions that may have wiped out many of the megafauna that were adapted to Pleistocene conditions.

Instead of perpetuating the traditional view that dogs are the products of our ancestors' ingenuity, Schleidt and Shalter (2003) argue that initial relationships between humans and wolves were cooperative and mutualistic. When we say wolves evolved into dogs, this means that first their behavior and ecology changed, and subsequently some changes were observed in anatomical features, such as body size, skull shape, and limb structure. One of the most intriguing aspects of Schleidt and Shalter's scenario is that wolves had a powerful influence on the social ethics of early human groups through a kind of "lupification" of human behavior or, as they put it, "humanization of the ape" (58). As medium-sized group-living carnivores, wolves survive by cooperation and teamwork. They hunt together, raise pups together, share food, and share risks. The pack system was successful not because of specific life history traits, anatomical features, or aspects of physiology or behavior. Its success derived from an entire array of specific features that made communal action and group survival possible.

In contrast, the social life of chimpanzees, our closest biological rela-tive, is opportunistic, almost Machiavellian in nature, with chimps seeking ways to get the better of each other (Wrangham and Peterson 1996). Schleidt and Shalter's (2003) idea emphasizes humans' coopera-tive side rather than their aggressive, competitive side. If we look for the biological roots of those traits we value as being especially humane, like group loyalty and cooperation for the common good, we don't find them in our ape relatives but in social canids like *Canis lupus* and *Lycaon pictus*.

As humankind spread across Eurasia, the sociality of human groups stopped resembling the primate model and began reflecting elements of the canid model (Schleidt and Shalter 2003). Based on paleontological, biogeographical, and genetic evidence, Schleidt and Shalter argue that this shift coincided with the time humans came into close contact with wolves after moving into their domain. Wolves, top predators, are also nature's first "pastoralists," living off migrating herds of ungulates such as reindeer, horses, and bison. During the last Ice Age, human groups adopted the wolves' pastoralist lifestyle and their group behavior (learning to cooperate, share risks, extend relationships beyond immediate kinship, etc.). This was the period during which earlier varieties of *Homo*—Neanderthals, Denisovans, and possibly east Asian *Homo erectus*—gave way to modern humans, and dogs began separating from wolves, as both species became more cooperative in their interactions, both within and between species.

A crucial trait of highly social species is the ability of members of social groups to accommodate one another rather than resorting to combat to resolve conflicts (Schleidt and Shalter 2003). It is not that aggressive conflicts never occur, simply that members of the group are constantly aware of where other group members are and what they are doing. All bonds are not the same and all interactions are not altruistic; there can be selfish behavior mixed in with cooperation. Marshall Thomas points out that individuals living and/or working together don't automatically have strong bonds or quickly establish rank, but nonetheless they generally prefer to avoid serious conflicts. Some tasks require full group participa-tion, such as driving off intruders or hunting large prey, and this means that relationships are amiable and ensure group survival (1993, 48).

If two species cooperate, this does not require that they become inti-mates outside of cooperative hunting. Some, like ravens and wolves, show aspects of intimacy (e.g., interspecies play), but each species can exist on its own, and all members of each species need not have the same relationship to one another. This is clearly shown by Australian Aboriginal peoples and dingoes (chapter 6). This is not surprising; after all, we have very different relationships with cats than we do with dogs.

One way to consider initial interactions between modern humans and wolves is that the two species encountered each other, shared kills, and sometimes may even have hunted together without any real affection or social bonds being involved. In some ways, this mirrors the relationships many rural people have with cats today. The relationship we have with our barn cat, described in chapter 9, was completely amiable and we were happy to see one another, but when our cat was hunting voles, she wanted nothing to do with us. We have had other feral cats, some of whom shared our barn for years, that we never were able to touch, even though we put out food for them in winter.

There was certainly a period, which may have lasted for many generations, during which humans and wolves learned about one another and made accommodations. As an example, consider the following fictional description from an Anishinaabe author of an encounter between a wolf and an Anishinaabe man contemplating suicide:

> I spoke to the wolf, asking my own question: "Wolf" I said,
> "Your people are hunted from the air and poisoned on the
> ground and killed on sight . . . stuffed in cages and almost
> wiped out. How is it that you go on living with such sorrow?
> How do you go on without turning around and destroying
> yourselves, as so many of us Anishinaabeg have done . . .?"
> The wolf answered, not in words, but with a continuation of his
> stare. "We live because we live." He did not ask questions. He
> did not give reasons. And I understood him then. Wolves accept
> the life they are given. They do not look around them and wish
> for a different life, or shorten their lives resenting the humans,
> or even fear them any more than is appropriate. They are
> efficient. They deal with what they encounter and then go on.
> Minute by minute. One day to the next. (Erdrich 2005, 120–21)

This is an imagined interaction, of course, but it rings true in the sense that a nonhuman cannot afford to indulge in self-pity or grief the way humans do. One factor we must keep in mind about our ancestors as they moved out of Africa and tried to solve the difficulties of life in Pleistocene Eurasia and North America is that they had to live like the wolf in Erdrich's account. These "were not people motivated by fear" (Derr 2011, 78), nor did they regard themselves as prey; they considered themselves effective predators (Pierotti 2011a, 2011b). Their life spans were not the "threescore and ten" that contemporary humans assume. Instead, like wolves, most died young and often from injuries sustained during hunting. Unlike contemporary humans, they did not expect to have multiple canid companions during their lifetimes; they may have known only one or at most two generations of a wolf family that could have inhabited that area for hundreds if not thousands of years. Wolf packs on Baffin and Ellesmere Islands have used dens thousands of years old and worn paths in stone that are ancient. "One doesn't often see very old things made by animals. [Looking at a path] I realized that I was looking at one of them" (Marshall Thomas 1993, 33).

Interacting with these successful, cooperative, and friendly (or at least not unfriendly) nonhuman neighbors, and given their incredible behavioral flexibility, our ancestors may have chosen to emulate wolves, especially in regard to hunting tactics and social dynamics. After all, "to come back unscathed [from a hunt], especially with a full stomach, is in itself a success story, and may explain why wolves seem to want mundane ordinary lives when not hunting" (Marshall Thomas 1993, 41). Our ancestors, faced with similar environmental conditions, may well have opted for a similar resolution and chosen to become the hunting and living companions of these experienced and social carnivores.

CHAPTER FOUR

Wolves, Archaeologists, and the Origin of Dogs

IT IS BEYOND THE scope of this book to review all archaeological and paleontological findings relating to wolves or dogs because this is a huge topic. Fortunately, recognized experts in this field have written insightfully about archaeological findings as they relate to the transition from wolves to dogs. We review and analyze the findings of these authors within the context of our idea concerning the processes by which early humans established relationships with wolves and how these might have led to the origins of what we consider to be domestic dogs. As we argued in the introduction, the initial establishment of these relationships is linked to, but decoupled from, the human-driven selective processes that more recently have produced significant phenotypic changes in wolves.

We initially discuss a classic archaeological case that may provide some interesting support for our idea of a long-term special relationship between humans and wolves, and also some insight into the nature of how archaeological findings might be interpreted. At Le Grotte du Lazaret, a 125,000-year-old complex of Paleolithic shelters in Nice, France, all larger shelters had skulls of wolves in their entrances (Thurston 1996; Franklin 2009). The age of this complex would push the origins of this relationship much further back than the dates generally accepted for wolf/human interactions. It does, however, give additional credence to the date of over 100,000 YBP suggested by Vilà et al. 1997 for possible origins of domestication, or at least associations between humans with wolves. This case further raises the possibility that members of the genus *Homo* other than *H. sapiens* may have had some sort of special relationship with wolves, because the date is well before modern humans arrived in Europe. This example was supposedly discounted when prominent archaeologist Lewis Binford concluded that the wolf skulls came from a "nearby wolf den, which accidentally intruded on the site" and had nothing to do with a relationship between *Homo sapiens* and *Canis lupus* (Thurston 1996, 14).

Such an argument, however, provides possible evidence for exactly the kind of scenario we proposed in the introduction—that a nearby wolf den, whose inhabitants were important to the humans, is likely to have been the sort of location where the first tentative relationships were developed. In regard to the idea that wolf skulls, not skeletons, reflect natural deaths, wolves rarely die around dens, and if they do, they do not leave only the skull. A 700-year-old den site on Ellesmere Island, where Marshall Thomas (2000) described ancient pathways created by wolves, has no skulls around it, despite centuries of continuous use (Mech 1995). Wolves, especially adults, are usually killed in action far from dens, and they would not continue to use a den if numerous individuals had died in that location. Binford, who seems to know little about wolves, their social dynamics, or their relationship with humans (1980), could not have been expected to have seen the possible significance of this association.

It is important to keep in mind the specific nature of archaeology as a science, particularly its history of conflicts with Indigenous peoples around the world and how this relates to our work. Such conflicts arise because archaeology involves highly disruptive research activities, often viewed by Indigenous peoples as disrespectful and even destructive of both physical landscape and ecological relationships that have been in place for long enough that they "exist on the other side of memory" (Marshall 1995, 207). This is a coded way of saying two things: (1) to Indigenous peoples there are some things about which knowledge is not needed, or perhaps should never be known; and (2) the relationship that these human communities have established with their homelands and dwelling places over millennia are disrupted by research activities integral to archaeological practices. Cultural traditions that are spatially rather than temporally oriented (Deloria 1992; Pierotti 2011a, 2011b) assume that ancient people who inhabited the same space are their relatives, and the resting places of relatives should not be disturbed, especially if the purpose is simply to increase knowledge in ways they consider unimportant (Thomas 2000; Pierotti 2011a, 2011b).

Indigenous peoples' beliefs concerning relatedness are expressed through interactions with nonhumans. These beliefs impact the study of the domestication of wolves in ways not readily perceived by investigators of European ancestry, nor would these attitudes manifest themselves through obvious anatomical changes in the wolves. Wolves are considered powerful and reliable allies by Indigenous peoples, whereas dogs are considered loyal but powerless companions because they do not control their own lives (Wallace and Hoebel 1948; Buller 1983; Marshall 1995; Pierotti 2011a; Fogg, Howe, and Pierotti 2015). Archaeology, however, depends on physical evidence, especially skeletal remains; therefore, if canid remains from a site appear to be those of a "wolf" rather than a "dog" according to statistical measurements (Boudadi-Maligne and Escarguel 2014; Morey 2014; Drake, Coquerelle, and Colombeau 2015; Germonpré et al. 2015), it is often argued that this could not be an animal associated with humans (Crockford and Kuzmin 2012; Boudadi-Maligne and Escarguel 2014; Drake, Coquerelle, and Colombeau 2015). Attitudes and perceptions of Indigenous peoples are not generally regarded as substantial evidence equivalent in importance to physical material found by scholars of European ancestry, even when examination of these attitudes might resolve possible conundrums in interpretation (Pierotti 2011a).

THE DEBATE OVER TIMING

As biologists, we find debates among anthropologists to be interesting but also frustrating. Evolutionary scholar Richard Dawkins quotes an anthropologist as saying, "The beauty of anthropology is that, when two anthropologists look at the same data, they come to opposite conclusions" (2015, 170). An example of this phenomenon can be seen in debates among archaeologists concerning the timing of establishment of relationships between humans and wolves. There seem to be two basic schools, one represented by Mietje Germonpré of the Belgian Institute of Natural Sciences and her colleagues, along with Patricia Shipman of Pennsylvania State University, that argues for the beginnings of this relationship occurring shortly after humans moved into the colder parts of Eurasia some

35,000 YBP (Germonpré et al. 2009, 2012; Shipman 2011, 2014, 2015). A contrasting point of view argues that animals identified as "Paleolithic dogs" (Germonpré et al. 2009) are in fact, unusual phenotypes of wolves, and that actual dogs did not appear until they were completely phenotypically distinguishable from wild wolves, primarily on the basis of a decline in overall size (Crockford and Kuzmin 2012; Boudadi-Maligne and Escarguel 2014; Morey 2014; Drake, Coquerelle, and Colombeau 2015; Morey and Jeger 2015).

What generates much of this ongoing debate is failure to recognize that two separate questions are being asked. The first is: Did *Homo sapiens* establish cooperative relationships with nondomestic canids, that is, wolves, shortly after moving into Eurasia? The second is: When and where did these nondomestic canids change into animals that could obviously be considered domesticated, that is, dogs? These questions emerge from arguments we made in the introduction concerning different components of "domestication" (see chapter 9 for expansion of these arguments).

One problem with trying to determine when wolves reached a point in their relationship with humans where they should be regarded as domestic organisms separate from wolves—dogs—is that this process almost certainly occurred in multiple places at various times in prehistory. One of the first scholars to make this point was Morey (1994), who argued that different wolf subspecies gave rise to dogs in different areas of the world. Subsequent findings based upon mitochondrial DNA sequences from archaeological dog remains from both Latin America and Alaska led Leonard et al. (2002) to argue that New World dogs arose from multiple Old World lineages that accompanied people across the Bering Strait land bridge, and that "it is not clear that modern dogs can be used to infer patterns of diversity in the past. . . . The number of domestication events for dogs thus remains unresolved" (cited in Morey 2010, 55–56).

The issue of when humans first had canid companions identifiable as dogs has recently had an interesting dimension added that supports the "early" school. Dillehay et al. (2015) present evidence that humans were in southern South America at least 18,500 YBP, which suggests that they must have been in North America much earlier. It is generally argued by the "late" school that when humans crossed from Siberia into North America they were accompanied by "dogs" (see Leonard et al. 2002).

Leonard et al.'s argument is based on the assumption that humans did not enter the Americas until the interglacial period around 12,000–13,000 YBP. The findings of Dillehay et al. 2015 indicate that if humans actually entered the Americas accompanied by dogs, this was more likely to have been at least 20,000 years ago and possibly earlier, suggesting that humans had to have animals recognizable as dogs well before the 14,000–16,000-year time frame argued for by the "late" school (Leonard et al. 2002; Morey 2010, 2014; Crockford and Kuzmin 2012; Boudadi-Maligne and Escarguel 2014; Drake, Coquerelle, and Colombeau 2015). The findings of Dillehay et al. thus provide support for the "early" domestication school of argument advocated by Germonpré and Shipman (Germonpré et al. 2009, 2012; Shipman 2011, 2014, 2015).

Examination of archaeological records suggests that around the world multiple groups of people had close relationships with wolves that over varying lengths of time led to domestication and eventually to animals recognized as dogs (Morey 2010). Also important to keep in mind is that, for the last 20,000 years, according to Indigenous accounts, canids living and working with many Native American groups were almost certainly wolves not clearly distinguishable from their "domestic" descendants.

Archaeological data are limited in its ability to date sites where humans are found with animals, combined with limitations in identifying specimens as wolves or dogs based solely on anatomical evidence. Studying archaeological canid remains found associated with humans, archaeologists employ only a few traits to determine if the animal was a "wild" (or predomesticated) wolf or a "dog." In the framework we propose, these two categories represent basically the same thing. For most scholars of this topic, however, this is where the debate centers: where and when this transition from wolf to dog occurs (Morey 2010, 2014; Boudadi-Maligne and Escarguel 2014; Drake, Coquerelle, and Colombeau 2015; Morey and Jeger 2015).

Morey is one of the most sophisticated archaeologists investigating canid domestication from an evolutionary perspective. In framing his arguments, he presents a wide variety of definitions as a part of a review concerning exactly what domestication means to various scholars (chapter 4 in Morey 2010). He acknowledges that the "domestication process" can involve an interaction between nonhumans and their environment, or even between two nonhuman entities (60–62), which leads him to criticize the concept of "artificial," that is, human-driven, selection.

Morey states, "The term domestication connotes a kind of symbiotic ecological relationship between two organisms, one in which they are closely involved in each other's life cycles . . . , irrespective of the particular physiological changes that may result" (Morey 2010, cited in Morey and Jeger 2015, 425). This is a fine definition, fitting well with the arguments we have been developing because it emphasizes process rather than outcomes. When considering canid specimens, however, Morey seems to abandon the concept that domestication is "irrespective of physiological changes," insisting that what identifies a domesticated "dog" is decline in size combined with "proportional reduction in length of the snout region, entailing a more steeply rising forehead area as well" (Morey and Jeger 2015, 424). The same individual who argued that "there is a probability that different wolf subspecies gave rise to dogs in different areas of the world" (Morey 1994, 345) and included a section titled "Which Wolf or Wolves?" in his major work (2010, 19–29) now seems focused upon an argument based solely on outcomes and ignoring processes, contending that dogs evolved in Europe. Morey and Jeger state, "Extinction of the particular wolf lineage that gave rise to dogs was a genuine possibility" (2015, 425), which exemplifies the sort of language associated with those arguing for a single ancestor in one particular location for all dogs, a position Morey has not previously taken.

This controversy arose because archaeological specimens consisting only of skeletal material make it difficult to distinguish large "dogs" from "wolves" (Morey 2010, 2014; Germonpré et al. 2015) and, as we argued in the introduction, "the relationship (leading to domestication) began as coevolutionary, with the species cooperating at times but also capable of functioning independently of one another. This state of affairs dominated the relationship between the two species during the early stages and may have persisted for 20,000 years or longer." Traits used to separate presumed "dogs" from "wolves" typically involve reduction in size, a shorter snout region and associated facial parts of the skull, and crowded teeth, some of which were positioned obliquely in the tooth row (Morey 2010), which are all relative characteristics. Some scholars differ in their views of exactly how far along these traits need to have progressed before a specimen can be considered a dog rather than a wolf.

The contrast between these ways of thinking is well illustrated in an important paper, "Human Domestication" (Leach 2003). Leach argues for "unconscious self-selection" and proposes that the environment, which includes interaction with other species, is an important factor in shaping evolutionary change. Despite this, however, the evidence used to assess such domestication is based primarily on assessing the status of fifteen possible outcomes. Five of these outcomes are skeletal (including body size, sexual dimorphism, and facial shape and structure), and the remaining ten involve "only soft tissues, biochemistry, and/or behaviour (which are) therefore archaeologically invisible in any direct form" (349).

This issue of "archaeological invisibility" is crucial because it means that archaeological arguments are predisposed to ignoring process, especially processes involving behavior and ecology. Germonpré and colleagues (2009) compared skull measurements of contemporary wolves; specimens from eleven contemporary breeds of large dog, including Belgian sheepdogs and central Asian shepherds, which are often left out of DNA-based analyses; and five well-recognized prehistoric domestic dogs dated from 22,000 YBP to around 10,000 YBP, subjecting these measurements to principal component analysis (PCA). These samples were subsequently compared to eleven canid skulls found associated with prehistoric human sites in Europe. One skull in particular, from Goyet Cave in Belgium, appeared to be that of a dog; however, it was dissimilar to contemporary dogs, falling between wolves and well-accepted prehistoric dogs in PCA space (figure 2 in Germonpré et al. 2009). This is what might be expected from an animal in the early stages of transformation from wild wolf to dog, that is, a very early stage of the domestication process, undergoing the coevolutionary process we described above. Two other specimens from Ukraine were also identified as early dogs, seven specimens appeared to be wolves, and the remaining specimen fit no available category (Germonpré et al. 2009). Of further interest to our arguments is that among the contemporary dogs examined, the central Asian shepherd was a clear outlier compared to the other dogs, suggesting these are not typical contemporary dogs (see chapter 5).

The most startling aspect of the specimen from Goyet Cave was its antiquity, dating back to 32,000–36,000 YBP, well before most scholars assumed the process of domestication arose. "The Paleolithic dogs in our data set are quite uniform in skull size and shape" (Germonpré et al.

2009, 482), suggesting one of two possibilities: (1) these specimens represent a unique form of wolf, or (2) early stages of wolf-human coevolution generated a consistent phenotype only slightly changed from wild ancestors through selective pressures rather than human-driven breeding. In terms of our arguments, however, these alternatives represent two sides of the same coin because we assume that "dogs" in early stages of domestication might be considered "wolves."

The question of how to identify a specimen as a wolf and not a dog appears controversial. One way to approach this issue is to employ the following reasoning: "All available information concerning domestication has in fact shown it to entail clear changes in the morphology, physiology and behaviour of the domesticated animal. . . . In archaeological contexts, the most common criteria for discerning domesticated from wild specimens derive from morphometric data and, to a more debatable degree, the presence of skeletal or dental pathologies and modifications. . . . Among these changes, the most significant are tied to an overall reduction in the size [of] individuals, more specifically, facial aspects of the cranium . . . and the retention of juvenile characters in domesticated adult canids" (Boudadi-Maligne and Escarguel 2014, 80). This argument has internal logic, but it is clear that this approach works best when attempting to evaluate human-directed aspects of domestication, when phenotypic changes are so obvious it is clear that we are no longer observing subtle changes that might result from coevolutionary pressures. In addition, the emphasis on overall reduction in size is contrary to our idea that during the early stages of wolf/human interaction, humans valued wolves for their ability to function as cooperative hunters and placed no selective pressure on these animals that would have led to a reduction in size.

A different critical view of the existence of "Paleolithic dogs" makes a similar point. An article supportive of the general interpretation yet critical of the methodology employed by Boudadi-Maligne and Escarguel argues: "Bivariate plots of linear distances and PCA of cranial ratios demonstrate that there is nearly complete overlap of dogs and wolves making any diagnostic criterion of phenotypic differences impossible. Because these measurements and analyses are insufficient for detecting morphological differences between modern dogs and wolves they should not be used when classifying fossil specimens. Boudadi-Maligne and

Escarguel's analysis included only very small archaeological dogs such that the large Goyet skull was not compared to dogs of a similar size. . . . There is no separation of dogs and wolves in a comparison of palate width to total skull length unless only very small dogs are included" (Drake, Coquerelle, and Colombeau 2015, 2). These scholars argue for the use of "3D geometric morphometric analyses to compare the cranial morphology of Goyet and Eliseevichi [specimens] to that of ancient and modern dogs and wolves." Despite methodological disputes, Drake, Coquerelle, and Colombeau conclude, "These Paleolithic canids are definitively wolves and not dogs" (1).

This is not a problem from our perspective; we have always assumed that during the early stages of this coevolutionary relationship the animals involved might be identified as wolves. In fact, we know from personal experience that the confusion described by Drake, Coquerelle, and Colombeau exists: when an X-ray of an obvious dog was sent to an American archaeologist, he identified it as a wolf according to PCA (see chapter 10). The problem was the archaeologist had not seen the animal in question whereas Pierotti had, and recognized that in coat color, pelage, behavior, and size it was clearly a dog and not a wild wolf. Drake, Coquerelle, and Colombeau also discuss results of ancient DNA analysis by Thalmann et al. (2013): "Analysis of complete mitochondrial genomes revealed that Goyet, and other Paleolithic wolves, belong to a sister clade to all ancient and modern dogs" (2015, 5). These results do not preclude these animals being the ancestors of at least some forms of dog (Shipman 2015). Drake and colleagues are firm adherents of the Coppinger view of dog domestication, endorsing the "wolves as scavengers" model of dog evolution. As a result, they fail to even consider the possibility that humans were interacting with animals that might be actual functioning nondomestic wolves.

Of the individuals questioning the existence of "Paleolithic dogs," the most sophisticated from an evolutionary perspective is Morey (2010, 2014; Morey and Jeger 2015), who now seems to assume that the boundary can be clearly distinguished between wolf and dog by focusing solely on outcomes rather then processes. Such ambiguity might result because Morey feels that animals undergoing domestication are still wolves, which fits well with our arguments as well as with contemporary systematics—dogs are simply wolves that have undergone a recent process of domestication. Morey argues, "Tellingly, the cases that they [Germonpré et al. 2009] suggest represent dogs show no appreciable size reduction relative to wolves" (2010, 28). This places a great deal of emphasis on the assumption stated by Boudadi-Maligne and Escarguel (2014) concerning size reduction, a point Morey emphasizes: "To be sure, most early dogs were systematically smaller than large, northern wolves. . . . Other domestic animals often exhibit similar changes in comparison to their wild ancestors" (2010, 20). As do Boudadi-Maligne and Escarguel, Morey seems to be limiting examples to obvious domestic dog specimens that are much more recent and small—less than 10,000 years old. This rule may not apply to much older examples of the process and does not really fit the definition of the domestication process he argued for in 2010 and repeats in 2015.

Reduction in size of domestic animals, mentioned by both Morey 2010 and Boudadi-Maligne and Escarguel 2014, is a commonly invoked outcome (J. Clutton-Brock 1981, 22; Leach 2003). Actual data on domestic animals are much more variable, however, and depends upon the role the specific type of animal assumes within human society (Zeder 2006). Among domestic ungulates, cattle, yaks, sheep, and goats are smaller than their wild ancestors; however, contemporary horses and pigs are typically much larger than their wild ancestors. The issue under investigation in this case, however, is the pattern that occurred in wolves as they changed into dogs. If humans depended on wolves as fellow hunters (Fogg, Howe, and Pierotti 2015; Shipman 2015), there should be little or no selection for diminished size; wolves that functioned naturally are probably what Paleolithic humans desired rather than smaller "village" dogs of the sort invoked by Morey (2010; Morey and Jeger 2015), Boudadi-Maligne and Escarguel (2014), and the Coppingers (2001).

There is little doubt that most (but not all) domestic dogs during the agricultural period of modern human evolution were small. The issue being addressed by Germonpré et al., however, concerns the appearance of early stages of wolf domestication:

We have shown that during the Pleistocene two sympatric canid morphotypes can be distinguished among the canid remains from several Upper Palaeolithic Eurasian sites. . . . One canid morphotype is most similar to recent wolves and is named by us

"Pleistocene wolf," while the second morphotype is distinct
from extant wolves. Relative to wolves, specimens from the
latter morphotype are characterised by short skull lengths, short
snouts, wide palates and braincases, and short mandibles as
well as small lower carnassials. . . . Both the upper and lower
carnassials of this latter morphotype are larger than those of
modern dogs. It is this morphotype that we described as
Palaeolithic dogs . . . to distinguish them from modern dogs. . . .
[These] can, but need not be the direct ancestors of recent dogs.
(2014, 211)

This interpretation fits well within the context of our conception
of early-stage interactions between humans and wolves. Genome
sequencing with recalibrated mutation rates from a wolf bone around
35,000 years old from the Taimyr Peninsula in northern Siberia
(Skoglund et al. 2015) leads to the inference that the split between wolves
and the earliest dog ancestors is consistent with the case for ancestors
existing 27,000–40,000 years ago.

As wolves adjusted to coexisting with humans, they might well have
undergone minor changes of their facial physiognomy to produce the
shorter, wider skulls described by Germonpré and colleagues (2014)
without significant reduction in size. These animals might still function
as wolves but could also be undergoing the domestication process as
defined by Morey above: "symbiotic ecological relationship between two
organisms, one in which they are closely involved in each other's life
cycles." Slight changes in facial structure might make wolf facial expressions easier for humans to read, which suggests that behaviorally these
animals were experiencing coevolutionary selection pressure to accommodate an increasing need for communication with humans, even if they
show no decline in size. The apparent inconsistency between Morey's
definition of domestication and what his criteria seem to be for "domestic
dogs" creates confusion. Changes such as declines in size would only
appear well after changes in behavior such as establishing "symbiotic
ecological relationships" through cooperation with another species in an
effort to increase rates of prey capture in a harsh and variable environment. The earliest individuals might actually increase in size because of
better food supplies (Shipman 2014, 2015).

FINDINGS FROM ANCIENT DNA

Ancient DNA analysis of specimens examined by Germonpré's research group was included in Thalmann et al. 2013 (discussed briefly in chapter 1) and invoked by Boudadi-Maligne and Escarguel 2014. The more striking result is that mitochondrial DNA from the Belgian specimens does not closely match mtDNA from any of the four clades of recent dogs identified by Thalmann et al., but seems to represent an ancestral group. Their haplotypes have not been found in either more recent fossil or contemporary dogs, nor have they been found in any wolves examined so far. Thus, they represent a sister clade to both contemporary domestic dogs and a selection of wolves from all over the world (see figure 1 in Thalmann et al. 2013; figure 12.2 in Shipman 2015). As pointed out in chapter 1, Thalmann et al. 2013 argue that contemporary dogs have their ancestry in Europe, although even a cursory examination of their figure 1 reveals four distinct origins of separate "dog" clades emerging from different groups of wolves, arguing strongly against a single origin. Most wolves in Thalmann et al.'s analysis are from Europe; however, they also include wolves from the Middle East (Iran, Saudi Arabia, Israel), India, China, Mongolia, and North America (Mexico and Alaska are separately represented as well). Some of their results make no biogeographical sense—for example, wolves from Israel and North America and wolves from Ukraine and North America are represented as each other's sister taxa (closest relatives). Our take on figure 1 in Thalmann et al. is that it represents the sort of jumbled interpretation that could be presented as evidence of why mtDNA results are of questionable use in assessing canid phylogenies (Coppinger, Spector, and Miller 2010). The four clades of "domestic dogs" shown in Thalmann et al. bear little or no resemblance to the domestic dog phylogeny shown in vonHoldt et al. 2010. This is perplexing, considering that Robert Wayne of UCLA is the "senior" coauthor (last listed) on both these papers.

Wayne's interpretation of the status of the Belgian specimens described by Germonpré et al. 2009 is that they are "wolves" rather than dogs because, as he states, "Morphologically they were diagnosed as dogs, but if they are dogs, they should be directly ancestral to modern dogs. We know they are not because their mtDNA sequences are outside the radiation of dog and wolf sequences" (quoted in Shipman 2015, 174). Again, this illustrates a problem with overreliance on DNA evidence. Shipman insightfully points out, "By this reasoning these canids are not wolves either, because their DNA does not fall within the known range of wolf sequences" (174–75).

A more serious issue is that Wayne does not acknowledge that even if these specific lines did not prove ancestral to contemporary dogs, this does not mean that they could not have represented a line of canids that may have changed in response to interactions with humans prior to going extinct. Such animals could have provided a model by which humans learned to coexist with wolves. These situations could have produced animals that subsequently gave rise to some lineages of domesticated wolves. Wayne's statement shows clear bias toward the idea that "domestic dogs" have a single ancestor, although the results of several papers of which he is a coauthor show obvious polyphyly, for example, Vilà et al. 1997; Leonard et al. 2002; Thalmann et al. 2013. He also fails to acknowledge the capacity of humans to learn from previous experience. Simply because the Belgian and Ukrainian specimens are all that we have at present does not mean that numerous similar situations could not have been occurring throughout Europe and other parts of the world where humans established close relationships with local wolves.

A more recent publication that incorporates evidence from both archaeological findings and ancient DNA makes it clear that dogs have at the least a dual origin, in both eastern and western Eurasia (Frantz et al. 2016). This work is unusual in that it also included DNA evidence from dog breeds not usually included in such studies, such as Saarloos wolf-dogs, Czechoslovakian wolf-dogs, and Tibetan mastiffs, finding that these breeds group primarily with other spitz-type dogs. Interestingly enough, Frantz et al. also include German shepherds among spitz-type dogs, a situation not previously reported in DNA studies but that conforms to the evidence presented elsewhere in this volume that German shepherds have some wolf ancestry and that they have been crossed with wolves to produce new breeds such as the Saarloos and Czechwolf. The difference in the position assigned to German shepherds in Frantz et al. and other studies (e.g., vonHoldt et al. 2010) suggests that this breed may be polyphyletic and not a true breed as identified by the AKC because different investigators appear to be working with different genotypes (see chapter 11 for more on this breed). More unusual is that in the neighbor joining tree generated by Frantz et al., the Saarloos wolf-dog, which is known to have resulted from a cross between a male German shepherd and a female wolf, with the offspring backcrossed to produce a consistent breed in the 1930s (Morris 2001), is identified as an outlier sister taxon to all of the other dogs. This makes no sense from an evolutionary perspective but suggests that there is something odd going on with dog DNA studies (Coppinger, Spector, and Miller et al. 2010; Pierotti 2014).

To return to the issue of the "Palaeolithic dogs" as described by Germonpré et al. (2009, 2012) these animals are somewhat larger but overall similar in shape to the essentially contemporaneous "incipient dog" from Razboinichya Cave in the Altai Mountains of Siberia (Ovodov et al. 2011). "The Razboinichya specimen is smaller than the smallest Palaeolithic 'dog' reported by Germonpré et al. (2012, Table 2, specimen 'Mezherich 4493') and significantly smaller than the mean for the group" (Crockford and Kuzmin 2012, 2798). In this case we have a canid that is small in size, yet according to its mtDNA, seems to be linked to wolves (Thalmann et al. 2013), which shows that changes in size can occur in wolves without their becoming dogs, at least from a genetic perspective. This specimen was found well away from Europe in a site dated to more than 30,000 YBP, which fits with our scenario, since we make no assumptions about size other than it should not be the major criterion for identifying wolves that lived in association with humans.

MODERN HUMANS, NEANDERTHALS, AND THE ROLE OF WOLVES

Some of these "symbiotic ecological relationships" may have eventually led to obvious domestic dogs acknowledged by Morey (2010) and the Coppingers (2001), while other lines remained indistinguishable from wolves phenotypically and genotypically, with some lines in this latter category going extinct. Shipman states: "Sometime in the range of 36,000 to 26,000 years ago, humans may have succeeded in breeding a wolf-dog from wolf stock, but . . . these . . . were not directly ancestral to modern dogs. . . . Though the Belgian DNA is not found in modern dogs, this does not rule out the possibility that this group was ancestral to modern dogs" (2015, 175). Shipman makes this argument as part of her examination of the dynamics of the relationship between *Homo sapiens* and *H. neanderthalensis* (2014, 2015). Shipman relies on archaeological records to reconstruct the history of Neanderthal people and their disappearance from Europe and the Middle East, where they had been present from about 300,000 YBP until perhaps 25,000 YBP. According to all accounts, Neanderthals were good hunters and physically stronger than *H. sapiens,* so there is some mystery to why they disappeared, or even how *H. sapiens* seems to have first displaced and then subsequently replaced them.

One way in which *sapiens* interacted with *neanderthalensis* involved interbreeding. Even though this was initially denied, this result is now well established through genetic studies (Krings et al. 1997; Mason and Short 2011; Sankararaman et al. 2012; Harris 2015). It also appears that *sapiens* interbred with Denisovans, another member of the genus *Homo,* almost exclusively known only from its DNA (Gibbons 2011; Harris 2015). DNA studies reveal that all humans other than Africans carry around 2–3 percent Neanderthal DNA, and humans from Southeast Asia, Indonesia, Australia, and New Guinea contain around 5–6 percent Denisovan DNA as well (Harris 2015).

More important from our perspective, however, is Shipman's (2015) contention that the reason *sapiens* replaced *neanderthalensis* (and perhaps Denisovans as well) is because of the advantage that the presence of "dogs" provided to *sapiens.* For these events to transpire, according to Shipman's scenario, humans had to have begun associating with wolves around 35,000 YBP, around the time the Neanderthals declined and finally vanished. Shipman endorses the findings of Germonpré et al. 2009; Germonpré, Lázničková-Galetová, and Sablin 2012; and Germonpré et al. 2013, which provide evidence that "dogs" came into being shortly after *sapiens* arrived in Europe. Shipman argues that a group of "ancient wolves with a separate mtDNA haplotype" could have evolved "if a group of closely related females sharing the same unusual mtDNA moved into a new area with their offspring and few outsider males" (2015, 170). This scenario is functionally identical to the model with which we opened the book, only we did not speculate about, or anticipate, unique DNA haplotypes.

Shipman points out, as we have, that it seems clear that humans involved in the first domestication, which undoubtedly involved wolves, were hunter-gatherers, refuting the scenario of wolves scavenging from human refuse dumps sometime in the last 12,000 years (Coppinger and Coppinger 2001; Drake, Coquerelle, and Colombeau 2015). Shipman further describes a strong association between early wolf-dogs and mammoth megasites, where human tools and other remains are found alongside the remains of multiple mammoths. She contends that this association implies a breakthrough in hunting techniques, and because there does not seem to be any evidence of new weapons or technology connected with these kill sites, she proposes, "This odd group of wolf-dogs was indeed a first attempt at domestication, [and provided] the advance that underpinned the formation of these mammoth megasites." This change in hunting techniques would have increased availability of food for both humans and wolves (dogs), which "fueled the continued growth of human populations and the range they were able to occupy" (2015, 181). This population increase is assumed to have led to increased competition for limited resources, eventually causing the extinction of Neanderthals. Increased food availability could have allowed wolves to maintain large size, countering the idea that decrease in size was a necessary outcome of human-wolf dynamics.

Mammoths represented far more than food. Their bones and tusks were useful for constructing shelters in steppe habitats where trees were not abundant (Shipman 2015). In addition, smaller bones and pieces of ivory would have been useful for making tools. Humans and their canid companions might not have killed mammoths, but may have been good at arriving shortly after mammoths expired from other causes, and the two species working together were capable of repelling efforts by other potential scavenging species, such as bears, hyenas, and even big cats (Shipman 2015). In a group, wolves are capable of standing up to grizzly bears, cougars, and probably any other predators with which wolves and humans coexisted. This would be particularly important if mammoths were dying in large numbers because of nutritional stress from changing environmental conditions, especially if hunting by humans, augmented by wolf companions, increased the stress on mammoth social groups (183–84).

Shipman thinks it important that the "dogs" identified by Germonpré's research group were large and powerful, countering criticism that these animals were not "dogs" because they did not show decrease in size (Morey 2010; Boudadi-Maligne and Escarguel 2014). Another point contradicting arguments that these Paleolithic canids were "wolves" is that isotopic analysis of the overall fauna from the Czech site where some "wolf-dogs" studied by Germonpré et al. were found revealed that specimens identified as obvious wolves ate horse and mammoth, whereas wolf-dogs seemed to rely more heavily on reindeer (Bocherens et al. 2014). This suggests that the "wolf-dogs" occupied an ecological niche different than that of the "wild" wolves, a point not addressed in any critique. Whatever the "wolf-dogs" may have been, they do not appear to have been typical wild wolves of their time.

We differ from Shipman on a few points. She seems to feel that wild wolves are too aggressive and dangerous to function as companions to humans (2015, 195). In our experience, wolves are calm and nonaggressive toward humans, and large dogs can be much more dangerous. These perceptions are reinforced by experiences of Indigenous Americans described in chapter 7. Shipman's thoughts on this converge on similar arguments from Francis (2015), cited in our introduction, that "all wolf-human interactions were overtly hostile." Shipman does not think of wolves as having initiated the relationship, instead relying on the Siberian fox study as a model, a case in which humans were in complete control and the setting was artificial (chapter 13 in Shipman 2015). She states, "I cannot envision the domestication of dogs happening in any but an accidental way, through trial and error that probably started out with a wolf pup or two and which only occasionally turned out well" (201). She also claims, "There is no way to estimate how long it might have taken early modern humans to domesticate wolves into wolf dogs or how many times keeping a wolf puppy around failed and ended in the death of the puppy" (200). This assumes that the Coppingers (2001) "Pinocchio hypothesis," that is, stealing and rearing wolf pups, was the only way in which domestication could have taken place, and also that humans must have controlled the relationship for it to work to their benefit.

Despite clever, insightful arguments, Shipman does not consider the possibility that wolves themselves may have initiated the relationship, a possibility that Morey (2010) does acknowledge. Her model cited above positing that "a group of closely related females sharing the same unusual mtDNA moved into a new area" assumes that adult females were involved because wolf puppies do not move on their own. Based on our experience, if the wolf is willing, it takes almost no time to establish the socialization process (chapter 11 and our conclusion). Gordon Smith, who probably spent more time interacting with wolves than any other person of recent European ancestry, states, "Wolves are friendly, gregarious creatures . . . quite willing to bond with any gentle, stable, warm-blooded creature, even man" (1978, 8). The Alaskan wolf known as Romeo shows that such individuals continue to exist, even up to the present day (Jans 2015). If Shipman would consider that wolves and humans were not simply competitors, and that wild wolves are not necessarily vicious, aggressive, or dangerous, her conceptual approach would be vastly strengthened.

MORE RECENT "DOMESTICATED" DOGS

Morey also seems partial to the "Pinocchio hypothesis," although he has argued, "Both [humans and wolves] were social species that hunted for the same prey items. Wolves . . . may have learned to be aware of human hunting activities and to scavenge from human kills. Perhaps humans even learned to do the same with wolves" (Morey 1990, quoted in Morey 2010, 69), which fits his concept of "symbiotic ecological relationships." If you assume the two species are learning to effectively use the other's skill sets, why not simply assume that adults could cooperate? Traditions of many Native American peoples, as suggested by Morey, discuss how in the early days of humans in the Americas, humans regularly scavenged wolf kills, and at times humans, especially women, seem to have been deliberately fed by wolves (chapter 7).

Morey and Shipman fall too easily back on ideas that humans and wolves were competitors with basically hostile relationships (sensu Francis 2015). As we pointed out in chapter 2, however, 85–95 percent of behavior among nonhuman species is now recognized to be cooperative rather than competitive. Morey acknowledges that the "deeply conditioned fear" resulting from the "big bad wolf" of Western fairy tales and folklore is "not likely to have characterized hunting-gathering peoples from thousands of years ago" (2010, 69) and cites the example of Alaskan (Inupiat) people who "do not express misplaced fear of wolves, but general admiration for their intelligence, sociality, prowess as hunters, and purposiveness and individuality of their behavior" (70). This is inconsistent with his invocation of the Coppingers' 2001 "scavenging wolf hypothesis" as a model for domestication, which he combines with Crockford's (2006) arguments about stress tolerance (Morey and Jeger 2015, 423). This may be a consequence of coming from a scientific tradition (archaeology) where humans are front and center in everything considered to be important. In archaeology, there is little emphasis on seeing the world from the perspective of other species.

Morey is a top scholar of the archaeology and recent history of obviously domestic dogs (1994, 2010). His primary research involves the circumstances surrounding burials of dogs, especially when they are interred alongside humans. There is one clear example of an animal identified as a wolf buried in a ritual manner with human remains; however, this case dates to only about 6,000–7,000 YBP (Bazaliiskiy and Savelyev 2003). Morphological changes unfortunately do not show the behavioral modifications that probably drove the morphological changes to begin with (Morey 1994; Schwartz 1997).

We have discussed the oldest possible dog fossil found, known as the "Goyet dog," dated to more than 30,000 YBP (Germonpré et al. 2009), and some of the reservations concerning this specimen. Morey is more comfortable with a considerably younger (14,000 YBP) mixed-species burial from a site known as Bonn-Oberkassel, where a dog was found buried next to a man approximately fifty years of age and a woman between the ages of twenty and twenty-five (see Morey 2010, 24–27, 53–55 for details). People living at the Bonn-Oberkassel site were hunter-gatherers, although the animal buried with the humans was morphologically distinct from wolves. It is a significant specimen because its very existence at 14,000 YBP contradicts the Coppingers' (2001) argument regarding the timing of domestication.

Large canid fossils were found on the central Russian plain at a site known as Eliseevichi and dated at approximately 13,900 years old (Morey 2010). These animals were larger than most gray wolves, closer in size to a Great Dane. Although the skulls were larger than those of most wolves, the arrangement of molars was more similar to wolves (Wang and Tedford 2008; Morey 2010). Size rendered the status of these specimens as either dog or wolf debatable, with the primary evidence for these animals as dogs being their unusual size combined with wide palates and short snouts (Morey 2010). "Alternatively, they might have been local wolves living in captivity, in close contact with humans, the descendants of an even larger wolf subspecies, or hybrids of some sort" (Miklósi 2007, quoted in Morey 2010, 26). After struggling to classify these remains, Morey concludes that "for many thousands of years after ca. 32,000 BP dogs showed no appreciable size reduction but then, remarkably, sometime after about 15,000 years ago, they rapidly did" (27), which seems to contradict the arguments in Morey and Jeger 2015.

The issue that seems to generate the greatest difficulty is that when examining archaeological findings, it is not possible to see when wolves began behaving socially in their interactions with humans, which is more important than when they began to look different than wolves from a physical perspective. All we can determine using such data are whether the humans showed high regard for the nonhumans, not if reciprocal relationships were in place. What evidence there is concerning behavior emerges from traditional stories of Indigenous peoples. These stories do not always distinguish between wolves and dogs, probably because the people themselves did not make such distinctions; therefore, if these stories specifically refer to wolves—that is, canids that did not coexist closely with humans—this is an indication that the culture in question considered its companions to be nondomestic, even if they shared social or ecological relationships. In other cases, it is clear that cultures distin guished between domestic and nondomestic canids, or at least between the animals they associated with prior to the arrival of Europeans and the domestic animals that they interacted with after that ecological and social invasion had taken place.

CONCLUSIONS

As we pointed out near the beginning of this chapter, there are two separate questions being asked: (1) Did Homo sapiens establish cooperative relationships with nondomestic canids, that is, wolves, shortly after moving into Eurasia? and (2) When and where did these nondomestic canids change into animals that could obviously be considered domesticated, that is, dogs? Problems arise when some investigators collapse these two distinct questions into a third that combines or conflates them: When did humans establish relationships with animals that can be easily recognized as domestic dogs? This is what generates much of the confusion that fuels the ongoing debate.

It appears that Germonpré and her research colleagues, along with Shipman and some others, for example, a group working in Russia (Ovodov et al. 2011; Druzhkova et al. 2013), are primarily interested in the first question about origins. In contrast, Ray Coppinger, Darcy Morey, and other critics of Germonpré are primarily concerned with the second question about obvious domestication; they frame their work, however, in terms of the third (combined) question. A clear example can be seen in

Morey's definition quoted above: "Domestication connotes a kind of symbiotic ecological relationship between two organisms, one in which they are closely involved in each other's life cycles . . . , irrespective of the particular physiological changes that may result." This definition is appropriate for the first question; however, it seems largely irrelevant to the second because it is designed to deal with how such relationships become established.

Another factor contributing to the overall confusion is that the DNA evidence should provide insight into the origins question; however, DNA researchers frame their results in terms of the second (domestication) and third (combined) questions. This situation seems to keep Robert Wayne and his collaborators in a constant swirl of apparently contradictory arguments, as is clear in the example discussed above in which, according to what Wayne told Shipman, the specimens discussed by Germonpré's group are not dogs but neither are they wolves according to his criteria. Interestingly, genetic data seem to invariably include wolves and dogs mixed together and suggest that dogs appear quite early as distinct lineages among wolves: for example, Vilà et al. 1997; Druzhkova et al. 2013; and Thalmann et al. 2013, all of which are papers on which Wayne is one of numerous coauthors. Unfortunately, Wayne also provides information such as figure 1.1.2 in Wayne 2010 and figure 1 in vonHoldt et al. 2010 showing dogs as a separate lineage distinct from all wolves. These results cannot be compatible with each other (or with the most recent results of Frantz et al. 2016), which is why the Coppinger group (Coppinger and Coppinger 2001; Coppinger, Spector, and Miller 2010) is highly critical of the results that Wayne presents. Most of this confusion seems to arise because Wayne and his colleagues seem, for the most part, to be determined to try to force fit contemporary domestic dogs into a monophyletic lineage, which contradicts evidence that Wayne and other collaborators present in other works (e.g., Vilà et al. 1997; Leonard et al. 2002; T. N. Anderson et al. 2009; Thalmann et al. 2013).

Our take on all of this is to point out that in many ways all of these important scholars are in basic agreement on key elements: humans and wolves established close, even symbiotic cooperative relationships, which gave rise to diverse lineages, some of which died out while others became what we describe today as domestic dogs. Shipman, and Germonpré and her colleagues, focus their efforts on early stages of this process (the origin question above), whereas Morey, Crockford, and others seem to concentrate on later stages that led to obvious morphological changes. In this context, all of their results are interesting and important. It would be helpful if we could acknowledge that the process of establishing relationships between humans and wolves happened many times in many places even up to the present day, and thus debates about exactly when and where are exercises in chasing our own tails (Pierotti 2014). The expert in human genetics Svante Paabo has been quoted as saying, "I'm often surprised how much scientists fight in paleontology. . . . I suppose the reason is that paleontology is a rather data-poor science. There are probably more paleontologists than there are important fossils" (Kean 2012, 212). This clearly seems to be the case in archaeology as well, where more than a dozen scientists are involved in deep debate over fewer than five fossils that seem to represent all of the possible specimens of dogs more than 20,000 years old.

CHAPTER FIVE

Asia

THE FIRST OF THE DOG-MEN AND JAPANESE DOG-WOLVES

ONE INTERESTING GEOGRAPHICAL REGION in which to examine relationships between humans and wolves is Asia, especially central Asia and Siberia, where the combination of harsh climate and rugged land-scapes makes it difficult for any but the hardiest beings to survive. In this area wolves and dogs have proved hard to distinguish, both genetically (Vilà et al. 1997) and morphologically. Central Asian dogs are very different than most forms of dog and more similar to contemporary wolves than in any other part of the world (Beregovoy 2001; Germonpré et al. 2009; Ovodov et al. 2011). Siberia and central Asia remain among the wildest, least populated (by humans) parts of the world. This isolation has generated numerous legends concerning the region's human inhabitants and their relationship with wolves or wolflike dogs.

Western civilization has long-standing traditions of dehumanizing other peoples, especially potential rivals for territory or resources during periods of colonial expansion. These traditions underlie racism, xeno-phobia, and other unpleasant attitudes still extant in people of European and Euro-American descent. One manifestation is the historical custom of identifying humanlike "others" as monstrous beings. In contemporary culture this attitude manifests itself in obsession with werewolves, vampires, and zombies. Werewolves and vampires were originally consid-ered by western Europeans to reside in eastern Europe. Locating monsters

FIGURE 5.1 Cynocephalus, 1493. (By Hartmann Schedel, Wikimedia)

in forests and mountains seems to be a result of the western European fear of nature and the need to destroy forests to promote "progress" (Sale 1991; Pierotti 2011a).

Werewolves are an interesting aspect of this bestiary (Summers 1966). The hybrid nature of such entities exemplifies one of the most consistent bugbears of "civilized societies"—humans who are part beast, sometimes called dog-men or cynocephali (White 1991). Such beings inspired a particular terror among post-agricultural humans because although they resembled humans, their identity was linked with wolves, now changed from objects of veneration and companionship into monsters (figure 5.1). Nameless, faceless dog-headed beings inhabit the same borderlands between fear and fascination as the vampires and zombies of contemporary cinema. According to ancient and medieval scholars, the location of these fearsome hordes was traditionally the far reaches of uncharted "wilderness," both physical and psychological (Pierotti 2011a).

This line of thought is exemplified in the description of the kynokephaloi by Herodotus, fifth-century BC Greek historian:

> It is said that there live in the mountains dog-headed men; they
> wear clothes made from animal skins and speak no language
> but bark like dogs and recognize one another by these sounds.
> Their teeth are stronger than those of dogs. . . .

> Kynokephaloi . . . do no work; they live by hunting, and
> when they have killed their game, they bake it in the sun;
> they also raise sheep, goats, and asses. . . . They drink the milk
> and sour milk of their sheep. . . . They dry fruits and make
> baskets. . . . Kynokephaloi have no houses, but live in caves. . . .
> Women bathe once a month, when they have their periods, and
> never at any other time. As for the men, they never bathe, but
> do wash their hands. . . . They have no beds, but make pallets
> of leaves. (White 1991, 49)

This description reads like accounts given by Europeans describing Indigenous peoples in Australia and North America, especially the part about "doing no work," as if hunting, gathering, and preserving food involved no labor. Indigenous peoples dried their meat ("bake it in the sun") to preserve it. Such attitudes reveal the Euro-bias of educated Europeans, starting with the ancient Greeks, who can avoid engaging in subsistence activities: hunting is sport; to do it for a living is to not *do work*. "These little dog-faced montagnards, with their amber, dried fruits, and tiny javelins, their sun-cooked meats, leafy beds, and rustic dress, and especially their great sense of justice . . . were Europe's original noble savages" (White 1991, 50).

Similar traditions and attitudes were also found among other "civilized" people. To people from India, China, and Europe, "uncharted" lands were the haunts of monstrous, barbarian races. Historiographic material from these traditions shows the Europeans placing cynocephali to the east; Indians placing "dog milkers and cookers" to the north and west, and the Chinese locating barbarian dog-men to their south and west (White 1991). The point of intersection for all three traditions is the interior of Asia, which is thus an ethnographic homeland for dog-men, suggesting that these stories involve Indigenous peoples of central and northern Asia. These accounts of societies of dog-men are juxtaposed with narratives about kingdoms of women or Amazons (White 1991). Many peoples from the Asian interior were polyandrous, with women playing a larger role in the culture than in the male-dominated traditions of Europe, India, and China (White 1991). Because these cultures were less hierarchical and more egalitarian, they were considered "less civilized." Central and western Siberia are also considered to be among the oldest centers of breeding and domestication of dogs, going back to the Paleolithic (Samar 2010), which is supported by discovery of a 33,000-year-old dog in the Altai region (Ovodov et al. 2011).

Altaic peoples of central Asia consider themselves descendants of a union between a male dog or wolf and a female human (White 1991), similar to stories told by another Altaic people, the Ainu of Japan (Walker 2005). An early Chinese source discussing "western barbarians" identified the Dog Jung (*jung* meaning wild, warlike, barbaric) as located to the northwest of Shang China (White 1991, 131). Ancestry myths of Altaian peoples, including Mongols, Tungus, Uighurs, and Ainu, may be one source of European myths of Amazon women and their cynocephalic male partners.

To Altaian peoples, the wolf was a "creature of great power and force" associated with their warrior societies. Wearing wolf or dog skins allowed warriors to "partake of the beast's raging spirit" (Golden 1997, 91–92). (In Norse traditions similar warriors were *ulfhednar,* or "those with wolf's head" (figure 5.2), who went into battle "without shields and were made as dogs or wolves" [92]). As with Native American tribes, there is confusion among outsiders about whether Altaian peoples thought of themselves as acting like dogs or as wolves, given that they may have considered "dogs" and "wolves" as basically the same entities. These hunting people, wearing animal skins, trading for cloth, flour, and metal, could be earlier examples encountered by Europeans who subsequently colonized North America and found the tribes of North America wearing battle clothing that made them resemble wolves (figure 5.3).

Accounts of wolflike males paired with Amazonian females might be metaphoric representations of cultural traditions in which men were hunters and warriors, spending most of the year away from their women, living with wolf or dog companions, while the women hunted small game, gathered, and maintained the camps from which the men would depart and to which they would return. Such a lifestyle probably resulted in high rates of male mortality, so women might well have paired with multiple men, including brothers, increasing the odds that each woman and her children would have a male provider of meat and skins, thus explaining the tendency toward polyandry as a mating system.

The Altai Mountains, which occupy an area where Russia, China, Kazakhstan, and Mongolia come together, are a culturally important

FIGURE 5.2 Pre-Christian warrior wearing a wolfskin. Such clothing could well have led to stories of dog-headed or dog-skinned warriors and hunters. (By KaOokami, Wikimedia)

landscape where much of early human prehistory seems to have come together. The Altai region of southern Siberia played a critical role in the peopling of northern Asia, serving as an entry point into Siberia. It has an old and rich history because humans have inhabited this area since the Paleolithic (Dulik et al. 2012). Local stories emphasize that Indigenous people of this area consider themselves to be among the world's oldest humans.

An important recent discovery from the Altai Mountains involves a cave called Denisova. Paleolithic, Neolithic, and later Turkic pastoralists took shelter in the cave, gathering their herds around them to wait out Siberian winters. Like the kynokephaloi described above, people here

FIGURE 5.3 Native American man wearing regalia consisting of a wolfskin cloak covering the head and upper part of the face. (Photograph by David Michael Kennedy, www.davidmichaelkennedy. com)

lived in caves and had wolflike dogs. "The cave's main chamber has a high, arched ceiling with a hole near the top that directs shimmering shafts of sunlight into the interior, so that the space feels holy, like a church" (Shreeve 2013). In July 2008, a young Russian archaeologist, Alexander Tsybankov, was digging in deposits 30,000 to 50,000 years old when he came upon a tiny piece of finger bone, which yielded sufficient DNA to provide a full sequence, establishing the existence of a previously unknown branch of the human family, the Denisovans (Shreve 2013). Denisovan DNA is recognized as part of the genetic heritage of modern humans from Southeast Asia (Harris 2015).

Not far away from Denisova Cave in the Altai Mountains lies Razboinichya Cave, where another specimen was found that is important to our story. This small canid skull dates to roughly the same time period as the Denisovans and appears to represent a very early stage of wolf domestication (Ovodov et al. 2011). This discovery and the Goyet specimen from Belgium (thousands of kilometers from the Siberian Altai)

described by Germonpré et al. (2009) are among the earliest "dog" specimens, providing further support for the idea that dog domestication was
multiregional and had no single place of origin. This may also be related
to the finding that some contemporary Russian wolves show mtDNA
"dog haplotypes" (Vilà et al. 1997; Coppinger and Coppinger 2001, 289).
What is interesting is that despite all the debate about the Goyet dog and
the fact that dogs have to be small to be identified as dogs (chapter 4), no
archaeologist seems to discuss the Razboinichya dog, which is decidedly
small as well as being over 30,000 years old.

Other archaeological data from this area reveal that local early
humans showed generalized and opportunistic hunting, with no single
prey species dominating. Species taken are those we might expect to be
the preferred prey of wolves, such as wild equids, wild sheep, reindeer,
and steppe bison (Goebel 2004). Wolves are the most common carnivore
species in these assemblages, but there is no evidence that wolves were
used as food. Altaic peoples traditionally employed small, mobile short-
term camps connected to large, more stable base camps, with hearths,
storage pits, and semi-permanent dwellings (Goebel 2004), which is to be
expected if the men, accompanied by wolflike companions, hunted over
large areas, while women, children, and the elderly maintained more
permanent communities. Such a social dynamic, which persisted into the
last few thousand years, probably led to stories of "dog-men" who were
connected to human, Amazonian women, because adult men in their
prime would only rarely be encountered in the permanent camps.

Another interesting specimen was found in 1995, when workers
building the Trans-Siberian railroad found a Neolithic cemetery at the
confluence of the Irkutsk and Angara Rivers southwest of Lake Baikal
(Bazaliiskiy and Savelyev 2003). In a single 7,300-year-old grave, archaeologists excavated the remains of a large wolf and at least two humans.
The wolf's remains were carefully arranged, with its head elevated, paws
placed against its body, and a human skull placed between its elbows and
knees. This is evidence of a socialized wolf living with humans (Bazaliiskiy
and Savelyev 2003) and suggests that in Asia between the Altai Mountains
and Lake Baikal over a period of at least 25,000 years, humans lived
closely with canids that might be either very primitive dogs or wolves.
Even today, shepherds in the Altai have wolflike laiki guarding their flocks
(Dr. C. A. Annett, personal communication).

LAIKI AND DOG-WOLVES

Since prehistoric times, wolves, or wolflike dogs called laiki, served native people as hunting companions and watchdogs. Wolves are sacred in the traditions of Indigenous Siberian peoples (Forsyth 1992). In the spiritual culture of the Nanai people of eastern Siberia, wolflike dogs serve as assistant spirits to shamans and as an intermediary between worlds (Samar 2010). Aboriginal laika types still remain with hunters of remote northern and northeastern provinces of the country (Cherkassov 1962). In the twentieth century, Russians saved hunting laiki from extinction, importing them from different geographic regions of Russia and breeding them to produce "pure" laika lines (Beregovoy 2001). Four "pure" breeds were established, of which two, the west Siberian laika and the east Siberian laika, bear the strongest resemblance to wolves. Their wolflike appearance, endurance, intelligence, and ability to survive rugged conditions with minimal care make them primitive dogs, although they can readily be socialized and become devoted to their human family (Beregovoy and Porter 2001).

Laika is the general term used for several types of hardy, powerful wolflike dogs that originated in Siberia (Beregovoy 2001). The Yenisei River divides Siberian laiki into eastern and western populations (Ioannesyan 1990; Beregovoy and Porter 2001). The recent rise of purebred dogs has come at the expense of the common all-purpose dogs of the people, whom they served for millennia. Although laiki can serve a number of functions assumed by large dogs in other parts of the world, we focus primarily on the laiki that are used for hunting. Laiki living with native peoples of the polar desert and tundra zones of Europe and Siberia became specialized for driving reindeer herds and/or pulling sleds; although some can become good hunting dogs, their capability is inferior to true hunting laiki of the taiga.

The term *laika* originates from the Russian *layat,* meaning "to bark" (Beregovoy 2001). The name simply means "dog that barks," although they also howl. Barking indicates their domestic status, along with tails that curl over their backs. Otherwise these dogs resemble wolves or wolf/dog crosses and would almost certainly be identified as wolf mixes by most of the American "experts" Pierotti has encountered. We briefly explore the history of the changing of the wolflike laika into one of the most popular "breeds" of dog in Russia.

Laiki are northern primitive breeds retaining traits of their recent ancestor, the wolf, in both appearance and behavior, but readily socialized to at least a few human companions. All have well-developed coarse outer coats and dense underfur, pointed muzzles, and pricked ears, which are wolflike features. In ancient Russia, there were as many laika "breeds" as Indigenous ethnic groups across the entirety of Siberia, and laiki may have originated from wolves multiple times (Beregovoy 2001, 2012).

Each laika lineage may be thousands of years old; nobody can be certain how each originally came into being (Beregovoy and Porter 2001). Some Russian scholars believe that laika ancestry can be traced back to the Stone Age, perhaps earlier. Nobody bred these animals in isolation, and they regularly interbred with wolves, although this is much more rare today (Cherkassov 1962; Beregovoy 2001). With some of these ancient lineages, the term *breed* may be inappropriate because they might represent direct descendants of geographic races of *Canis lupus,* unique phenotypes that evolved in specific localities, separate from other "dogs" (Sabaneev 1993; Beregovoy 2012). If these lines were bred directly from nondomestic *Canis lupus* in different areas, they may in fact be separate evolutionary lineages with separate origins, which would make them not examples of domestic dogs (it is certain that these forms were not among those used by Linnaeus to identify *Canis familiaris*) but separate evolutionary lineages with their own origins and possible endpoints.

We emphasize west Siberian laiki because of Pierotti's experience observing this breed in the Altai region of Siberia and also because this breed is featured in Herzog and Vasyukov's 2010 film, *The Happy People* (see below). West Siberian laiki, lightly built and fast, were established by selective sampling of aboriginal dogs of Mansi (Voguls) and Hanty (Ostyak) (figure 5.4) (Ioannesyan 1990; Beregovoy 2001).

Standard west Siberian laiki show these characteristics: body structure is strong; temperament is well-balanced but live and alert; height at the shoulder fifty-five to sixty centimeters (twenty-two to twenty-four inches); coat is well developed with thick undercoat. The body is well developed with deep and long chest (this is a wolf characteristic, not seen in many otherwise wolflike breeds, like huskies and malamutes, that are used as sled dogs). Coat color is predominately gray with white, pure white and with spots. Dogs of the Borka-Panda strain of the west Siberian laika are characterized by so-called *zverovatost* (a wolflike look), a peculiar

FIGURE 5.4 West Siberian laika. Note the overall wolflike appearance; the only non-wolflike feature is the tail curled over the back. (Courtesy of Vladimir Beregovoy)

primitive similarity to wild ancestors characteristic of laiki in general (Ioannesyan 1990; Beregovoy and Porter 2001).

Modern Russian laika breeders preferred bigger dogs; some were about sixty-two to sixty-three centimeters (about twenty-five inches) at the shoulder, and a few males were up to sixty-eight centimeters (twenty-seven inches) at the shoulder, making them comparable in size to wild wolves. Although the breed has more recently reverted to a smaller overall size, it remains bigger than aboriginal dogs of the Mansi and Hanty peoples (Ioannesyan 1990).

The pelage of the west Siberian laiki consists of a double coat of harsh straight guard hairs and thick and soft undercoat that keeps them warm even in Siberian winter (Beregovoy 2001). In winter, dogs in countries with a cold climate have hair growing between their toes. In dogs such as huskies and malamutes, dark colors typically have sharper boundaries than are found in wild wolves, in contrast to laiki, which have a wolflike coloration.

Female west Siberian laiki have traditionally had one estrus per year, usually in February–March, similar to that seen in wild wolves but not in most domestic dogs, which can have two or more estrus periods a year (Beregovoy 2001). The first estrus usually occurs between the age of one to two and a half years, which is also more similar to wolves than to contemporary domestic dogs, which typically have their first estrus at the age of six to eight months. Russian experts do not recommend breeding laiki until they are at least two years old, which is also the age of first reproduction in wolves (Beregovoy and Porter 2001; Haber and Holleman 2013). Female west Siberian laiki are good mothers and, if conditions permit, dig their own whelping dens. They raise their puppies without assistance if sufficient food is available. In recent years, after the collapse of the Soviet Union, breeders have selected for females that can have two estrus periods per year to increase the number of puppies and the resulting income (V. Beregovoy, personal communication).

BEHAVIOR IN RESPONSE TO HUMANS AND OTHER DOGS

West Siberian laiki are affectionate and devoted to their human companions. Attitude and behavior toward unfamiliar people varies individually among laiki, depending on the situation (V. Beregovoy, personal communication). Like captive wolves, many west Siberian laiki have traditionally been aloof with strange people, avoiding hands and watching them suspiciously, which is behavior typical of socialized wolves or dogs with recent wolf ancestry (see our chapters 10 and 11 and Beregovoy 2001). As with wolf-dog crosses, west Siberian laiki accept new owners with difficulty and need time to adjust to a new social situation.

West Siberian laiki are very inquisitive about animals, including wild animals, and all dogs have a strong hunting desire (Cherkassov 1962). The hunting behavior of west Siberian laiki is species specific, however, and serves to satisfy the needs of the human hunters with whom they associate (Voilochnikov and Voilochnikov 1982; Beregovoy 2001). Those developed by Mansi and Hanty Indigenous peoples live in close company with reindeer herds without attacking or killing them. In Russian villages and small towns, laiki are trained to ignore livestock, including cows, pigs, goats, and sheep. As with wolves, small animals such as cats, rabbits, and poultry are more tempting because they resemble natural prey, but laiki learn to leave cats living in the same household alone. Rabbits, however, seem irresistible, and must be kept in sturdy enclosures. West Siberian laiki can learn not to kill chickens, but even reliable dogs may backslide when relocated to a new place (Beregovoy 2001; see Fentress 1967 for such an example involving a pure wolf discussed in chapter 11).

FIGURE 5.5 West Siberian laika playing with mixed Alsatian puppy at Lake Teletskoye, Altai Republic, Russia. (Photograph by R. Pierotti)

Laiki and other dogs are treated very differently in Russia than are dogs in the United States and Canada. These animals seem to roam freely among people, even in cities as large as Moscow (Spotte 2012). Some have social ties to families, but others seem to wander on their own. Without detailed study, it is difficult to determine if an individual is a "pet" or a free-ranging animal. Female dogs that are ready to produce a litter will sometimes dig a den under a house or exposed tree roots (Spotte 2012).

Free-ranging dogs in many parts of the world live by scavenging and by predation on small mammals and livestock (Spotte 2012), and such animals appear to have provided the model for the ideas of Coppinger and Coppinger (2001), although it is important to emphasize that these animals are not like free-living wolves, either in social structure or ecology (Spotte 2012). In contrast, free-ranging dogs in Russia are often fed by humans directly and are well-accepted members of society. Some even ride the Moscow subway system, knowing when and where to get off (Spotte 2012). Pierotti observed dogs in Gorno-Altaisk in southern Siberia wandering around markets and other areas without any aggression on their part and no hostile reactions from humans. They made no attempts to steal food, but at times vendors or other people would offer them food, which was politely accepted. To see such animals, which in the United States would be identified as wolf-dogs and feared, peaceably wandering and interacting with both people and other dogs (figure 5.5), is reinforced by the statement of Marshall Thomas (2000) that Europeans handle their relationships with dogs much better than Americans. From a scientific perspective, such findings also provide a context for the quasi-hostility toward dogs found, for example, in the writings of the Coppingers.

HAPPY PEOPLE AND DOGS ALONG THE YENISEI

An example of a contemporary society in which the women run the settlements while the men are off hunting much of the year can be seen among Siberian people. Russian filmmaker Dmitry Vasyukov conducted years of work on Bakhta, a village in the geographic center of the Siberian taiga along the Yenisei River, occupied by 300 people sustained by hunting and trapping, especially for sable, *Martes zibellina*, the large weasel important in the Russian fur trade (Herzog and Vasyukov 2010). Today these hunters are ethnic Russians; however, in pre-Soviet times Indigenous Siberian peoples, especially the Evenky, filled these roles. The men of Bakhta spend considerable time alone in the woods trapping, accompanied only by their laiki. This social system reflects the millennia-old model we described above, in which men live as hunters and trappers, alone with their dogs for most of the year, while the women remain in permanent villages, tending gardens and livestock, caring for elders, and raising children. The men show their strongest emotional warmth when speaking about their dogs— not their children or wives, or even their country or their youth (Herzog and Vasyukov 2010). Vitally important for them is the right wolflike dog that will obey commands, assist in hunting and, if necessary, sacrifice its life for its human companion.

At least one laika is always at the hunter's side, as critical to his survival as a trap or a gun. Dogs sleep in the snow outside or in small huts built specially for them, are capable of running for miles at a time, and can be used to pull small sleds filled with hunting and trapping gear. When one trapper returns to Bakhta for New Year's Day by snowmobile along the frozen Yenisei, his laika runs alongside—for over ninety kilometers without food, demonstrating the incredible endurance and physical conditioning of these dogs.

Under extreme conditions, the relationship between a trapper and his dog is more than that of "best friends"—it is a life-and-death symbiosis. The bond between the animal and the man is the essence of this cultural tradition. The trappers feed only a small amount of food to their dogs in the morning. In *The Happy People* (Herzog and Vasyukov 2010), one trapper claims that in lean winters his loyal dog fed him rather than the other way around, which is basically what we argued for the origin of the wolf-human relationship in the introduction. The trappers have a collaborative relationship with their dogs to rustle out and kill small game, including sables.

Mark Derr (1995) refers to laiki as the "curs of the Taiga" because they remind him of the multipurpose dog, one with the intelligence and ability to do many different things and pursue a wide variety of quarry. Laiki may bay up a moose or brown bear or wild boar, then dive into a frigid river to fetch a shot duck (Cherkassov 1962). They may even dive after reindeer that are swimming in the river. The "happy" men of Bakhta featured in the film are alone in the wilderness from fall through early spring with only their dogs for company and support. Living in accordance with the rules and ethics established by nature, these men express contempt for greedy individuals among them who overfish, overhunt, and trap out of season for a few extra rubles. Laiki are the dogs of people who live off the land, and they represent sustenance, both material and emotional, to their male human companions (Cherkassov 1962).

The finding that some Russian wolves show dog mitochondrial haplotypes (Vilà et al. 1997) may result because female laiki used to readily breed with wolves, and some of their offspring may return to a wild lifestyle (although this is more rare of late because breeders are seeking a standard; V. Beregovoy, personal communication). Laiki are capable of hunting and feeding themselves, and as described above, they can sleep outside with minimal shelter in a Siberian winter. Laiki readily assume such roles because these trappers live for much of the year as contemporary hunter-gatherers.

Another aspect that must be considered is that humans trying to live alone would suffer from being solitary social animals. With one or more dogs or wolves as companions, however, humans survive better emotionally, in addition to the advantages of having multiple individuals to hunt together and for one another (Cherkassov 1962). Such individuals might become "dog-men" in the eyes of "civilized societies" in which most men don't catch, butcher, or prepare their food. Relationships between Siberian hunter-trappers and their laiki probably resemble those that Paleolithic humans had with the wolves that eventually became the ancestors of all dogs many thousands of years ago. The Siberian taiga has been dog country since the earliest days of the dog (Cherkassov 1962), back to the time of what we might call dog-wolves, which helps to place the discovery in Razboinichya Cave in proper context (Ovodov et al. 2011).

The Herzog and Vasyukov (2010) film shows how such deep bonds could have formed between wandering hunters and their dog-wolves in the initial stages of domestication (Schleidt 1998; Schleidt and Shalter 2003). As with free-living wolves, each laika is an individual with its own character and eccentricities, making it important for humans to develop personal relationships in order to achieve cooperation (V. Beregovoy, personal communication). Effective trappers rely upon the strength of the bond with their laiki rather than on "discipline" to maintain these relationships (Herzog and Vasyukov 2010). Dogs have their own ideas and stick with them, making them effective companions who might come up with unique solutions to problems that might otherwise overwhelm the human (Cherkassov 1962). An abusive owner may be abandoned by his companion, another factor that could explain the presence of "dog" mtDNA haplotypes in contemporary Russian wolves (Vilà et al. 1997).

Trappers treat great dogs like retired hunters, feeding and caring for them as long as they live in recognition of the bond and the service they have provided (Herzog and Vasyukov 2010). One trapper recalls holding his old bitch in his arms as she died after fighting a brown bear invading the village, standing her ground after other dogs had run. Hearing the commotion, the trapper grabbed his gun, ran to the scene, and fired at the bear, which was tearing up his dog while she ripped at it with her teeth. His shot grazed the bear's paw, and it charged him at high speed, so he shot it point-blank in the face, knocking it backward. Without even waiting to see whether it had gone down, the trapper grabbed his dog and raced for the nurse's station, trying to save her.

One point that warrants emphasis is that relationships between Indigenous peoples of Siberia and wolves can vary among cultures. In the shamanistic traditions of Indigenous Siberians, bears, wolves, and ravens are the primary totem animals (Czaplicka 1914). In cultures that have become pastoralist, however, and herd reindeer, yaks, sheep, or goats, wolves are seen as potential threats to livestock, and although this dynamic is not as hostile as the one in contemporary Euro-American society, wolves can be killed. One key trait of laiki is that they are trained not to attack the livestock, sometimes leading to situations involving physical conflict between wild wolves and their laika relatives over protection of livestock (C. A. Annett, personal communication).

JAPANESE WOLVES AND MOUNTAIN DOGS

In contrast to Siberia, Japan represents a very interesting situation with regard to the history of interactions between humans and wolves in Asia while also revealing the pernicious influence of Euro-bias shaping Western-style thinking and treatment of wolves even in a non-Western society. This is well documented in a monograph (2005) dealing with environmental history by Brett Walker of Montana State University, who used his familiarity with Japanese culture and history as well as his fluency in the Japanese language to explore documents normally not known or available to scholars in Europe and the Americas. Walker's findings illustrate potentially important differences between local knowledge and the traditional Western approaches to classification that reveal key insights concerning the nature of wild wolves, domestic dogs, and potential crosses between the two.

Japan has a history of rigorous scientific exploration that was largely independent of Western influences until the nineteenth century. From 1600 to 1800 natural scientists in Japan emphasized an Eastern rather than a Western worldview, and thus their approach was "folk-biological," according to Western thought, characterized by "a holistic appreciation of the local biota" (Walker 2005, 30). In practice this meant they relied on morphological and behavioral patterns recognized as commonsensical by most people within the culture rather than depending solely upon esoteric information gathered by "experts." Within such an approach, dogs and wolves were regarded as similar beings with the boundary between them not clearly defined, in contrast to Linnaean taxonomy.

The historical Japanese approach is close to what we propose in this book, in which we have made clear our belief that Linnaeus was incorrect in regarding these two forms of life as different species, even though this interpretation seems logical to most people with a "Western" orientation and manifests itself in American and European legal precepts to this day (chapter 10). That orientation also fuels arguments developed by contemporary scientists such as the Coppinger group (Coppinger and Coppinger 2001; Coppinger and Feinstein 2015) as well as debates about when a wolf becomes a dog that can be recognized by archaeologists.

The Japanese traditionally recognized two forms of wild canid as native to their main islands: *okami*, or wolf, and *yamainu*, mountain dog, even though it seems that for most of Japanese history there was a tendency to regard these as referring to the same type of creature (Walker 2005). Some Japanese have contended that the Chinese term *sai,* which translates into the Japanese *yamainu,* can also refer to animals in the genus *Cuon,* Asiatic wild dogs, also referred to as dholes, which are quite different genetically from wolves, having a different number of chromosomes. What is interesting in the Japanese perception of these two forms was that the okami was considered to be a benign and helpful creature, whereas yamainu were thought to be more aggressive and dangerous (Walker 2005). There was also a third form of canid, the much larger, Hokkaido wolf, found on the northernmost Japanese islands, which is sacred to the Indigenous Ainu, a Turkic people unrelated to the majority of Japanese (White 1991; Walker 2005).

The logic underlying these differences in perception is that for a culture that grew grain and hunted but did not keep domestic ungulates, okami was considered a protector of agricultural fields because the species preyed on the deer and wild pigs that were destructive to farmers' crops. Thus farmers encouraged the presence of wolves around their farms and fields, a practice anathema to most people of European descent (Walker 2005). In contrast, yamainu was considered to be a more aggressive animal, especially toward humans.

In addition, Japan has primitive wolflike indigenous domestic dog breeds: the large Akita, the medium-sized Kishu and Shikoku, and the small Shiba Inu. All are spitz-type, primitive breeds closely related to dingoes, according to some DNA evidence (vonHoldt et al. 2010). Some of these breeds have been developed to hunt in packs, and their ancient heritage means they can readily breed with wolves. In the nineteenth century, Western observers commented that Japanese dogs looked and acted more like wolves than the dogs with which they were familiar (Skabelund 2004). Edward Morse, an American zoologist at Tokyo University in the 1870s, characterized Japanese breeds as being of the "wolf variety" and described the behavior of packs of Japanese village dogs in the following manner: "At night they are very noisy, making sounds, like cats, but more infernal, they howl and squeal, but never bark" (1945, 52).

Japanese cultural attitudes made cross-breeding between wolves and dogs likely, and some specimens of yamainu have been identified as probable crosses between okami and domestic dogs (Walker 2005, 52). This is of interest because public health records demonstrate clearly that

some wolflike dog breeds are more dangerous to humans than are socialized or even wild wolves (see chapter 10). Japanese wolves were relatively
small for wild wolves, and Akitas may actually be larger than Japanese
wolves; this may have relevance to issues raised in chapter 4 about
whether wild wolves must show a decline in size to be considered
domesticated.

The general conclusion is that Japanese wolves, at least the friendlier
and more sociable okami, were nondomestic but were "wolves in the
process of sustained, if regionally specific, hybridization with dogs"
(Walker 2005, 54). Walker poses an interesting question: "If artificial
and natural selection both played a role in the evolution of Japan's wolves,
and the emergence of the *yamainu*, then what purpose do these terms
really serve [other] than to illustrate some last ditch attempts to distinguish humans, the creators of artifice, from the rest of the natural world?"
(55). This point is similar to one made by Ritvo (2010, 208): "As the
valence of the wild has increased and its definition has become assertion,
rather than description, the boundaries of domestication have also
blurred."

The situation was more clear-cut on Hokkaido, where the Ainu
people regarded the Hokkaido wolf as a "god" (Walker 2005). A more
subtle and accurate rendition may be found in Walker's statement that "it
was part of their ethnobiological understanding of the world that Ainu
realized that their hunting habits resembled those of the Hokkaido wolf,
and such recognition fostered reverence for the animal" (83). Euro-bias
often leads to assumptions that religious traditions of many cultures are
peopled by "gods" and that the people in question "worship" such entities,
even though to many of the peoples involved the practices and beliefs are
not similar to those found in Western monotheistic traditions (Pierotti
2011a). Recall that the Ainu are an Altaian people who believed that their
origin resulted from a white wolf mating with a goddess (White 1991).
One similarity with some Native American tribes is that the Ainu had a
tradition in which they would leave behind portions of kills for wolves. In
exchange, if they cleared their throats near a wolf kill, the wolf "god"
would generously make room for them and allow them first access to the
kill (see Schlesier 1987; Marshall 1995; Walker 2005; Fogg, Howe, and
Pierotti 2015 for similar examples in North America). Another similarity
between American tribes and the Ainu is that their word for "wolf," *setu,*

also means "dog" (see Fogg, Howe, and Pierotti 2015). Ainu saw "the two kinds of canine as similar and their distinction, when one was necessary, was largely situational. When in the village aiding people, the canines were dogs; when in the mountains hunting deer they were wolves." Ainu tried to reproduce or encourage "wolf traits in their own dogs . . . through both accidental and intentional breeding" (Walker 2005, 85). In at least one Ainu village they "tried domesticating wolves," which involved Ainu caring for wolf pups in their village for about two years (until adulthood). Once the wolves had become accustomed to people, the Ainu "allowed them . . . into the mountains alone to hunt and kill deer, after which the wolves returned to the villages" (86).

Walker quotes an epic Ainu poem that concludes with the following message:

Therefore,
simply put
a dog,
even if you kill one,
should not be sent in the direction
of the ocean.
Its ancestors are wolves.
It should be sent in the direction of the mountains.
That's the lesson of this story. (Walker 2005, 90)

For Hokkaido wolves, a different and lethal lesson was coming with the Americans who arrived in Japan during the nineteenth century (Walker 2005). American military personnel and diplomats came initially, followed quickly by traders and entrepreneurs. The Americans decided that the Japanese diet of vegetables, grains, and fish was insufficient and that the Japanese needed to raise domestic animals, especially cattle. The introduction of bovids meant that they had to be protected from potential predators. American hatred and fear of untamed nature would also be brought to a nation that had never previously shown such attitudes. U.S. range management practices were introduced, including the use of poisoned baits. Tragically, these genocidal practices were very effective, and thousands of years of peaceful coexistence between humans and Hokkaido wolves came to an end. Today wolves are apparently extinct across the entire Japanese archipelago. From the wolves' perspective, this event can be summarized in a haiku Pierotti wrote:

> Americans come
> Bringing their poison with them
> Wolves will be no more.

CHAPTER SIX

"Dingo Makes Us Human"

ABORIGINAL PEOPLES AND *CANIS LUPUS DINGO*

IN THE OTHER CASES we discuss, humans moved into lands occupied by wolves and had to work out relationships with these four-legged social hunters. There is one place in the world where the situation was reversed—humans were present well before the social canids arrived, creating a dynamic relationship unlike any other. *Homo sapiens* and *Canis lupus dingo*—the Aboriginal people of Australia and dingoes—were the only large placental mammals present on an entire continent.

Unlike other types of dog, dingoes are considered a true subspecies because they appear to have had only a single origin, almost certainly in southern Asia, independent of any other domestication events that may have been taking place in Europe, central Asia, or North America. Before they were introduced to Australia and New Guinea (New Guinea has a close relative, the New Guinea singing dog), dingoes appear to have been human companions that had undergone selection and morphologically diverged from wolves. Dingoes differ from wolves in their superficial anatomy, such as color, head shape, and pelage. If any wolf descendant should be considered a separate species, it would be the dingo; however, they remain interfertile with wolves, at least with domesticated wolves, that is, dogs. They are a semi-domesticated form capable of living on their own, even though they often maintain some sort of relationship with humans (Meggitt 1965). Until the British came to Australia in 1788, there were no domestic, or camp, dogs, and dingoes lived in loose association with humans.

The history of dingoes in Australia is closely tied to the relatively recent history of Aboriginal human groups because dingoes did not originate in Australia. They are the only nonvolant large placental mammals other than *Homo sapiens* among the otherwise almost exclusively marsupial terrestrial mammalian fauna of the continent (Corbett 1995; Rose 2000). The domestication of the wolves that were the ancestor of dingoes and similar dogs, for example, New Guinea singing dogs, also possibly Taiwanese mountain dogs and African basenji (Corbett 1995; Dayton 2003a, 2003b), is thought to have occurred somewhere in southern Asia. Many such dogs are found in Southeast Asia, New Guinea, the Philippines, Taiwan and, according to Corbett 1995, even the Japanese breeds described in chapter 5 are all "dingoes" according to some scholars.

Although separate domestication of wolves was likely to have occurred around the same time period, or even earlier, in central Asia, the Middle East, Europe, and North America (Morey 1994, 2010), the process was much different in southern Asia, where there seem to have been greater changes in coat color but little change in skeletal features because the humans with whom they associated were hunter-gatherers in a dry environment. In contrast, in western Asia and Europe over the last few thousand years, wolves were subjected to selection pressure that favored physical characteristics making the wolves beneficial to agricultural or pastoral humans in performing tasks such as guarding, hauling, or hunting gazelles, resulting in, for example, Afghans and salukis. This led to significant changes in the skeleton and musculature of some breeds, such as huskies and malamutes and even some Russian laiki, without altering the basic coat color and pattern of the wolf ancestor.

Most breeds of dog today have been around for less than 500 years; however, a few breeds such as Afghans, salukis, and various eastern Asian breeds, such as Akita, shar-pei, and Shiba Inu, may be a couple thousand years old. Proto-dingoes in southern Asia were probably kept for the same reasons that Native Americans began cooperative relationships with wolves; they were valued for their hunting skills and vigilance (but not as beasts of burden, given their body type). In Australia, dingoes also offered forms of spiritual or interspecies companionship that were not possible with the predominant marsupials (Rose 2000, 2011, personal communication).

Australian humans apparently made no effort to manipulate the physical structure of their animals through selective breeding: the dingo has "remained virtually unchanged in its morphology for at least 5500 years" (Corbett 1995, 14). Corbett argues that with southern Asian forms, little or no deliberate selection occurred, and therefore the Indian wolf, C. *lupus pallipes,* a relatively small and short-haired form, has moved into the contemporary world little changed, except perhaps in coloration. Contemporary dingoes share many of the same morphological characteristics found in early canid subfossil remains. According to accounts from Europeans in Australia, dingoes were used "for food, hunting, alerting and perhaps other cultural reasons" (13). Another perspective, based upon the observations of one who actually lived with Aboriginal people, argues that Aboriginal peoples used dingoes as companions, hunting assistants, warm bedfellows on cold nights, and watchdogs (Rose 2000).

The apparent lack of phenotypic change in Australia was probably linked to the different style of relationship Aboriginal peoples had with dingoes. When dingoes were introduced to Australia around 5,000–6,000 YBP, Aboriginal peoples accepted them as they were, feeling no need to attempt to domesticate them; thus, dingoes were no longer subject to any form of selection other than simply adapting to environmental conditions in Australia. In other parts of the world the combination of natural selection and some selective breeding altered the size and appearance of early canids while also producing more easily socialized ("tame") individuals to produce a genetically domesticated version of *Canis lupus* (Crockford 2006; Morey 2010).

Like wolves, dingoes also live in family groups; however, humans regularly influence this group structure. Pairs of dingoes have frequently been spotted hunting together, but nowadays evidence of hunting groups larger than two is uncommon, probably because there are no longer any large prey in Australia that require cooperative hunting. The following excerpt from a reminiscence about life in 1860s southwestern Australia describes how dingoes benefit from hunting in pairs:

> We were driving along a low ridge of hills and the valley below
> was clear and grassy, with a few tall trees from which the
> afternoon sun threw long shadows, when we caught sight of two
> wild dogs chasing a large kangaroo. The poor hunted brown

kangaroo hopped first one way, then another, while the two dogs, which were of beautiful golden sable colour, seemed to be acting on a settled plan. They kept heading the kangaroo off as it ran. At last it could hop no longer, so stood with its back to a big tree and tried to fight off its enemies with its feet, but they were too wary to go too close. One would lie down some distance away, while the other worried and snapped at the kangaroo. When it was tired, the one lying down took its place. . . . At last [one] dog caught the kangaroo off its guard, and made a spring at its neck; in a second the other dog rushed up and attacked on the other side, and the poor beast was pulled struggling to the ground, and in a few moments was dead. I think the combat lasted quite half an hour. (A. Y. Hassell, cited in Meggitt 1965, 12)

WILD, FERAL, OR DOMESTIC?

Dingoes may serve as a model for some of the earlier stages of transition from a wild to a domestic state that we discuss in chapter 9. Every place they occur, dingoes function as either wild or semi-wild animals—they live independently but associate with humans in many ways, including hunting and even sleeping together (Meggitt 1965; Rose 2000). The relationship between dingoes and Aboriginal peoples has been characterized as commensalism: "the process whereby two organisms live in close association but do not depend on each other for survival" (Corbett 1995, 12). This is important because it strongly suggests that domestication of canids is likely to have been a multistep process emerging from a relationship between wolves and humans and may not necessarily involve obvious attempts at domestication.

One major requirement for a form of organism to be considered domestic according to most definitions is the development of complete dependence on humans (Morey 2010; see chapter 9). Many domesticated species lose the ability to return to the wild and would perish without help from humans; pigs and horses have been known to successfully go feral in North America, but fully domestic dogs rarely survive long on their own (Spotte 2012; but see Terhune 1935 example in our introduction). Dogs also seem incapable of successfully breeding independently because males have lost the behaviors associated with parental care.

Dingoes do not follow this pattern; it is clear from their appearance and behavior that some phenotypic changes probably took place before they were introduced to Australia. Dingoes had to have been transported to New Guinea and Australia by humans in boats because there were no land connections that would have allowed dingoes to come to Australia on their own. Humans who brought dingoes to Australia bred them for some purpose, whether it was purely companionship, as rat catchers, or perhaps as a potential food source on long trips out at sea (Corbett 1995). Living in close proximity to these humans (on boats) would have required the dingoes to show little or no aggressive behavior. It is possible that the dingoes brought to Australia on boats were the tamest ones from the area where the humans originated. This might have resulted in an animal that was comfortable with humans and sought a relationship with them but was still capable of living independently under most circumstances. Today dingoes function largely as a wild species just like wolves, but they are capable of living and cooperating with humans.

RELATIONSHIPS BETWEEN ABORIGINALS AND CARNIVORES

One major issue in investigating and discussing the relationship Australian Aboriginal peoples had with the dingo is that the accounts by European observers vary widely. Meggitt (1965) described everything from abuse to loving relationships, from cooperative hunting to Aboriginal people following dingoes to steal their kills. Other sources claim that humans did not hunt with dingoes at all. It is difficult to determine if these differing accounts simply represent variation in relationships, or if there also exists the possibility of racism and paternalism (Euro-bias) on the part of the European invaders of Australia. Australian scholar Deborah Bird Rose, who has worked closely with Aboriginal people, describes the situation this way: "Whites love to talk about how Aboriginal people do things badly. I'm sure one can find instances of dogs being maltreated, and perhaps even of indifference to their being killed. In general, though, people love their dogs like they love the rest of the family—usually lovingly and kindly, some-times angrily, sometimes helplessly, and often in ways that are not identical to Anglo-European ways" (personal communication).

It is important to keep in mind the point we made at the beginning of this chapter: Australia was a land where humans existed for at least 40,000 years before dingoes arrived, and thus Aboriginal culture was already well established. The culture evolved in a land where the dominant mammals were marsupials, which had radiated to fill many ecological niches similar to those occupied by placental mammals in most parts of the world. This included two very unusual forms of what were essentially giant carnivorous possums. One of these was named *Thylacoleo carnifex,* or the "meat-eating lion with a pouch," by Richard Owen, the notoriously anti-Darwin but excellent nineteenth-century British comparative anatomist (1859b). Although the animal was described as lionlike, its physiognomy suggests it more closely resembled a very large hyena. Although *Thylacoleo* did have catlike paws, its front limbs were longer than its hind limbs, which would have made it relatively slow and not capable of catlike attacks. *Thylacoleo* weighed around 100–130 kilograms (220–300 pounds), which is more the size of a jaguar than a lion. It survived in Australia until around 30,000 years ago, so it was contemporaneous with humans for a minimum of several thousand years.

The other large marsupial carnivore was the thylacine, or "Tasmanian tiger," *Thylacinus cynocephalus,* the "pouched dog with a dog's head." Referred to as tigers because they were striped from the shoulders to the base of the tail, thylacines are well known to Western science because they survived until very recent times (the 1930s) in Tasmania, and on the Australian mainland until the last few thousand years (Guiler 1985; D. Owen 2003). Thylacines were nowhere near as impressive as *Thylacoleo,* being about the size of a small wolf (another common name was Tasmanian wolf), with adults being about 60 centimeters (24 inches) at the shoulder and weighing 20–30 kilograms (45–70 pounds). Both of these marsupial carnivores were relatively slow, especially when compared with most placental carnivores, like wolves or big cats.

Aboriginal traditions and stories feature large carnivores. One story tells of a "huge, doglike creature" called Tjularka that stalked from behind and leaped upon prey (Strehlow 1971, 140). Another Aboriginal tale regarding predators and human relationships, re-created from a story recorded by Aboriginal elder Dick Roughsey (D. B. Rose, personal communication), is called "The Giant Devil Dingo," although details suggest that it might not really be about actual dingoes. In this story Grasshopper Woman controls a giant devil dingo that shakes the ground when it walks. She uses this creature to hunt down humans for food. One day two young men come to speak with Grasshopper Woman. After unsuccessfully trying to persuade them to camp near her, she sends the giant devil dingo to hunt the men down. The men run for days trying to escape the giant devil dingo, but eventually they grow tired and have to rest. They hide on opposite sides of a ravine and wait for the giant devil dingo to pass by them. When it passes, the two spear the creature, killing it, then bring it back to their people to be eaten. The medicine man takes the bones, skin, and kidneys of the dingo to the top of the mountain, and from them he creates two smaller dingoes. He tells these dingoes they will not hunt humans—instead, they will be companions to the humans and assist them in their hunts for food (Roughsey 1973).

There is some question as to whether the giant devil dingo is actually meant to represent a literal dingo. Australian anthropologist Deborah Bird Rose suggests the following interpretation:

> The term *devil* is usually a gloss for something that has no exact counterpart in everyday life. In the [YouTube] video, the giant devil dog is called Kaya. The story has many resonances with similar stories across the continent. In the Victoria River country of the Northern Territory, Kaya are glossed as "devils": they take the form of human skeletons, and they are the custodians of the bones of the dead. They are not content to take care of bones, however, and actively prey on living humans. Their companion dogs are called Mulukurr, sometimes glossed as "devil dogs," and they work with the Kaya as they prey on living humans. . . . The Mulukurr was once a thylacine; . . . the term *devil* is added to make the point that there are no everyday creatures of this type. The term *giant* usually means large in size and/or significance. The stories from both regions [Cape York and the Victoria River region] show some of the tension in being both predator and prey, and articulate efforts to contain predation against humans. (D. B. Rose, personal communication)

There are some other curious aspects to the account of the giant devil dingo. First, the giant devil dingo is solitary, rather than living as part of a family group, the way actual dingoes do. Second, the giant devil dingo pursues the men for days, which suggests that it is quite slow and tracks them by scent. This suggests a third possible candidate for the giant devil dingo. Australia was also home to another, very large predator when Aboriginal people first arrived around 50,000 YBP. This was *Varanus priscus,* or megalania, a monitor lizard (goanna in Australia), large enough to have been a predator on its sister species the Komodo dragon (Head et al. 2009). The largest terrestrial lizard ever found, megalania was five to eight meters (fifteen to twenty-five feet) long and weighed well in excess of 500 kilograms (1,000 pounds) (R. Owen 1859a). This is a creature that could well have shaken the ground when it walked, and it also tracked prey by smell, or at least taste, since lizards use their tongues to detect odors and check scent trails. Deborah Rose, however, thinks it unlikely that Aboriginal stories would conflate reptiles and mammals (personal communication)

Another point is that Aboriginal people regularly use currently existing goannas for food, but we do not know if they ever ate the other mammalian predators; after the giant devil dingo is killed, it is eaten. It is also possible that the giant devil dingo represents a conflated memory of a couple, or even all, of these now-extinct predators. Megalania and *Thylacoleo* died out within a few thousand years after Aboriginal peoples arrived in Australia, and thylacines disappeared from the Australian mainland shortly after actual dingoes arrived, which may have been responsible for their disappearance (D. Owen 2003). The final point is that the medicine man takes the bones, skin, and kidneys (?) of the devil dingo and uses them to create a pair of smaller dingoes who are to become "companions to humans, and assist them in hunting." This is the sort of story found within the traditions of many Indigenous peoples, relating how a now-extinct form gave rise to a successor species that works with humans (see also chapter 7).

Rose is the author of the classic ethnographic study of Aboriginal cultures and spiritual systems, *Dingo Makes Us Human* (2000), whose title expresses her understanding of the significance of the relationship between the two species from the human perspective. Aboriginal people believe in shared characteristics of humans and dingoes, as shown in another creation story reported by Rose: "In Dreaming, only the dingo walked then as he does now. He was shaped like a dog, behaved like a dog, and dingo and human were one. It was the dingo who gave us our characteristic shape with respect to head and genitals. Dingoes are thought to still be very close to humans: they are what we would be if we were not what we are" (47). The emphasis on the shape of genitals refers to the fact that humans and dingoes were the only placental mammals, and thus had standard-issue mammalian genitalia. Marsupials have very unusual genitals; in some cases the scrotum is in front of the penis. It is crucial to emphasize that although Aboriginal people do not regard themselves as the owners of dingoes, the relationships were very personal—dingoes knew their human families, and the families knew their dingoes (D. B. Rose, personal communication). Dingoes are treated as mutually benefi-cial partners in this relationship and are respected as such. This is very similar to the way many Native American groups see their relationship with wolves.

As pointed out above, Australian Aboriginal peoples had no other placental companions until dingoes arrived around 5,000 YBP. Rose describes the relationship: "Aboriginal people love (and loved) these animals so deeply. . . . The presence of dingoes probably offered a new way of thinking about humans—as the creation stories suggest—how we are and are not dingoes. . . . I would imagine that before dingoes Australia was rather a lonely place. There were almost no placental mammals (only bats and mega-bats), and there were none that were social with humans the way dingoes are" (personal communication). In many Euro-Australian accounts it is acknowledged that their Aboriginal companions often dote upon dingoes: "Its master never strikes, but merely threatens it. He caresses it like a child, eats the fleas off it, and then kisses it on the snout" (Meggitt 1965, 16). The dingo, from this account, more closely resembles the wolf than the domesticated dog in several key behavioral traits, including group hunting. Another feature of dingoes, one that is also used to distinguish between wolves and domestic dogs, is that dingoes, like wolves, do not bark but howl and squeal (recall the description of Japanese dogs in chapter 5). Despite its obvious differences from most domestic forms, the dingo was classified as simply one form of domestic dog, *Canis familiaris dingo,* until the 1990s, when it officially became recognized as a subspecies of wolf (*C. lupus dingo*). This differentiates dingoes from the domestic dog, which cannot be a valid subspecies because of multiple origins and highly variable phenotypes (Pierotti 2012a, 2012b, 2014).

It is clear from the traditional stories what lasting impressions the dingo has made on the lives of the Aboriginal peoples. It is likely that Aboriginal people encouraged, or at least allowed, dingoes to live independently, because they have largely returned to the wild since their initial introduction. Humans and dingoes, however, both chose to maintain a commensal relationship. As was true of the Ainu people in Japan with wolves, individual dingoes could be taken from their dens after being weaned, then socialized by specific humans to the requirements of living in human society such that they became valued companions (Meggitt 1965; Rose 2000). Once they had been socialized, they were allowed to come and go as they pleased.

Numerous accounts exist of humans using dingoes while hunting, although the type of interaction varies (Meggitt 1965; Corbett 1995; Rose 2000). The Garawa people used dingoes to pursue wounded animals. "The dingoes would wait for the hunter to signal the prey had been hit and then would scent the blood and pursue and harass the animal until the hunter could catch up" (Pickering, cited in Corbett 1995, 18) Other accounts involving the Garawa people describe dingoes driving animals toward hunters or grabbing frilled lizards as they ran in front of the grass fires used to drive other animals in a certain direction.

Aboriginal peoples were also able to hunt at night more efficiently when accompanied by dingoes thanks to the latter's superior senses of olfaction and hearing. After dark dingoes could be released to roam the rocky hills, where they would make sounds to signal the prey had either been caught or "bailed up" (Corbett 1995). The Mildjingi tribe used dingoes to drive large goannas and native cats (*Dasyurus*, a housecat-sized carnivorous possum) into trees where they could be easily speared by hunters. These dingoes were also used to chase bandicoots into logs, from which they could be retrieved by hunters with axes. Organized parties including dingoes and humans were used to drive kangaroos into an ambush to be killed (Thomson 1949; Corbett 1995), and Aboriginal peoples also carry out ritualized accounts of battles between large kangaroos and dingoes (Strehlow 1971, 305).

Certain groups of Aboriginals prefer to hunt accompanied by wild dingoes because they are more effective in running down large game. The following account describes how this relationship between the dingo and the Aboriginal people began: "A group of Walbiri men out hunting picked up the relatively fresh tracks of a dingo . . . which in turn was trailing a kangaroo. If necessary they followed the dingo for the whole day, endeavoring to overtake it just as it was pulling down the exhausted quarry. Without harming the dingo, the men dispatched the kangaroo with spears and boomerang and gutted the carcass then and there preparatory to carrying it back to the camp. . . . It was common for . . . offal to be left for the dingo to eat" (Meggitt 1965, 19). It is clear that the relationship between the Aboriginal people and the dingo is a long-running, complex one characterized by mutual benefits accruing from successful mixed-species hunts. Dingoes are left with the part of the animal they would typically eat, and the humans are supplied with food for themselves and their families. This cooperation strengthens the bonds between the two species since both prosper because of it.

The importance of dingoes to Aboriginal peoples is clear from the fact that many of their origin stories center around dingoes. This relationship is illustrated on rock art found in the Australian desert depicting Aboriginal man and dingo standing side by side. The following dingo story originates from the Yamatji (Tjupan) nation around the Mount Magnet, Meekatharra, and Sandstone in the Murchison of Western Australia. They were told to Lorraine Barnard by her elders about the dingo, or Wunghoo, and recorded by Jeff Barnard for Nic Papalia. This story, "The Birth of the Dingo," shows the longtime respect the Aboriginal people have had for the dingo and their close relationship:

> Long ago in The Dreamtime there was an enormous black
> kangaroo. He would chase and kill people whenever he saw
> them hunting. . . . Many brave men tried to kill this kangaroo
> only to be slain. . . . One day the leader of the clan called a
> meeting. . . . He said . . . I have this magic axe . . . made of
> special stone. . . . I am giving this axe to my sons to collect
> wood from the sacred country of the mulga trees. From this
> wood, I will carve a guardian creature that will fight the monster
> kangaroo and protect us. . . . Spirit dingo was made with mulga
> branch for a backbone, forked sticks for ears, the teeth of a
> marsupial mole, and the tail of a bilby. . . . After several days and
> nights . . . the clever Mabarn man was finally able to bring the
> great big Dingo guardian to life. . . . The Dingo had enough
> power, hatred and venom to destroy the monster kangaroo. . . .
> Some clan members and the Mabarn man's sons set out to hunt
> and gather food, with the Dingo following closely behind

carefully disguised. Bringing small children was a clever ploy to flush the Monster out. . . . The hunting party came across the monster kangaroo fast asleep in the afternoon shade near a large breakaway hill. The sons shouted and woke the monster kangaroo from its slumber. . . . The dingo pounced and grabbed the monster kangaroo by its throat and killed him instantly. The hunting party and the Guardian returned home and the sons told their father what had happened. Now the people were able to hunt and gather food near the billabong without fear of the monster kangaroo killing them. To this day the elders had two or three dingoes as pets to protect them from harm and dingoes do not bark, they howl. (Watson 2016)

There are several important themes in this story. First, the importance of hunting is evident, particularly in more bountiful areas. The dingo made it possible to hunt in these better regions, which means that the relationship formed between the two species was mutually beneficial. Interestingly, recent paleontological work in Australia has actually discovered evidence of large carnivorous or omnivorous kangaroos, although these seem to have gone extinct long before any humans arrived in Australia (Wroe 1996; Wroe, Brammall, and Cooke 1998). It is always difficult to determine just what the "Dreamtime" means; however, there were large kangaroos capable of killing and eating other large vertebrates at some time in the past, so there might be some basis to this story in ancient memories.

WOLF AND DINGO

Theory concerning how and when the split between wolf and dingo took place focuses on the idea of multiple domestication events occurring in different parts of the world (Morey 1994, 2010; Pierotti 2012b, 2014). "This could include early Middle Eastern breeds like mastiffs and greyhounds, Scandinavian dogs, and dingoes, all as breeds that have been domesticated separately from other more common domestic dogs" (Gade 2002). There is debate regarding whether the dingoes that were brought to Australia from Asia were somewhat domesticated but then escaped and went feral, or whether they initially lived with people but were allowed by humans to live independently in Australia. The stories above tend to lend credence to the latter theory.

The oldest dingo fossils found in Australia are approximately 5,500 years old and closely resemble the skeletons of contemporary dingoes, in both size and physiognomy (Corbett 1995). Even in the very different environment of Australia, which is largely dry and more temperate than the areas of Southeast Asia and New Guinea (home of the dingolike New Guinea singing dog, *C. lupus hallstromi*), the basic morphology of the dingo remained very similar to the original introduced form. Dingoes do not look like modern dogs, but they are also very different in appearance from classic northern wolves, although they are similar in size to the smaller wolves of India (*C. lupus pallipes*). Pure dingoes come in four basic colors: ginger, black and tan, black, or white. In contrast, the coat color of a wolf may range from pure black to pure white, with any tint or shade of gray, tan, cream, ochre, sienna, and brown. Dingoes typically weigh around fifteen kilograms (thirty-five to forty pounds), which is about the size of a coyote, but there have been reports of larger dingoes in certain areas with large prey populations (Corbett 1995). Dingoes conform to the idea argued by Morey (2010) about wolves becoming smaller in size after undergoing at least some sort of domestication process.

Wolves and dingoes are more similar behaviorally than they are physically. The same reproduction rules exist within a pack of wolves and a pack of dingoes: only the alpha female and her mate are permitted to reproduce. In a wolf pack the alpha pair will try to prevent subordinates from reproducing by preventing copulation. In a pack of dingoes, subordinate members of the pack are permitted to mate and produce offspring, but the offspring are killed and eaten by the alpha female within days (Corbett 1995). Wolves and wolflike dogs will kill but not eat neonates produced by a subordinate female (McLeod 1990; Marshall Thomas 2000). The wolflike behavior of dingoes, as demonstrated in their ability to live separate from humans and survive after their initial domestication, has been retained in contemporary wild dingoes. Dingoes and New Guinea singing dogs are regarded as wild rather than domestic animals, even though some people keep such animals as (very independent) pets (Corbett 1995).

The only clear indication of some form of domestication in dingoes prior to their introduction to Australia can be found in their physical appearance. Without this early influence on dingo physiognomy by humans, dingoes probably would have remained more physically similar to Indian wolves. If early domestication altered only the appearance of the dingo, not the behavioral tendencies, then it indicates that other breeds of domestic dogs might also be capable of living separately from humans, as can be observed in some wolflike dogs such as Siberian laiki, Alaskan malamutes, Japanese Akita and Shiba Inu, Korean Jindo, and even border collies (Terhune 1935; see our introduction). This is not to say that any domestic dog could be released and it would successfully go feral, feeding itself and rearing young; but specific breeds that have been changed only in minor ways might still retain this ability to survive without humans as dingoes do.

Many breeds of dog that are used by Indigenous peoples in the north remain similar to wolves—for example, Alaskan malamutes, Siberian huskies, Eskimo dogs, and the spitz-like dogs from eastern Asia. These animals are often bred with wolves to retain their wolflike characteristics. Dingoes, on the other hand, have not been interbred with wolves for a very long time. In contemporary Australia, however, dingoes do interbreed with domestic dogs, the domestic version of wolves (Corbett 1995). The resulting offspring may be more comfortable living with humans or, alternatively, they may choose to return to the bush to live free of humans. Genetic research suggests greater similarity between domestic dogs and dingoes than between wolves and dingoes. Still, most scientists regard dingoes and their close relatives as a distinct subspecies of wolf, *Canis lupus dingo*, even though other domestic dogs should be considered as simply varieties (Pierotti 2012a, 2012b, 2014).

The relationship between Aboriginals and dingoes is best described as commensal as defined by Corbett 1995. In this situation the Aboriginal people are able to benefit from the relationship, while the dingo is left relatively unchanged by the situation. The Aborigines received a wide range of benefits from this relationship, including a cooperative hunting partner and/or a protector.

Aboriginal people believe the dingo was created to help and protect the people, a belief inspired by their close relationship (Rose 2000, 2011). Dingoes were often used to keep Aboriginal people warm on cold nights out in the bush; the popular term "three-dog night" referred to cold desert nights during which numerous dogs were needed to keep a human warm. The dingoes would pile up near the humans, sharing their body heat. Having a dingo around could mean the difference between life and death for Aboriginals under extreme conditions.

In contrast, the dingo appears to be mostly unaffected by the relationship. It is said that dingoes that live in the wild often look like they have been fed better than the ones kept in the camps with humans (T. L. Mitchell, cited in Meggitt 1965). When associated with humans these animals spend their days hunting for food for the humans to eat, and then their nights trying to feed themselves, which supports the idea that dingoes might actually be better off without humans, and could explain why over time many dingoes return to the bush. As puppies, dingoes are often raised among the Aboriginals to help with hunting; however, as they mature they become more interested in their own independence (Meggitt 1965). "The dingo often ran away, especially in the pairing season and at such times it never returns" (C. Lumholtz, cited in Meggitt 1965). "What evidence there is suggests that, as a rule, dingoes taken from the bush return to it sooner or later" (Meggitt 1965, 22).

This intermittent adoption or cohabitation resembles the relationship described between the Ainu and wolves on Hokkaido (chapter 5) and between Indigenous Americans and wolves, in which humans would visit wolf dens to play with puppies and mark them, probably so that friendlier puppies could be adopted once they were weaned (Fogg, Howe, and Pierotti 2015). There are two meanings of weaning. The first is going off milk and on to solid food, which is a good time to adopt a puppy. "Where the Seven Sisters go, dingo knowledge goes too. The Pleiades tell Old Tim's people that the dingo pups are being born, and when they make another shift they tell that the pups have opened their eyes. The old people, all those long gone generations of Aboriginal countrymen, would raid the dens, finding food and companions" (Rose 2011, 56).

The second meaning of weaning is the time when pups are able to hunt for themselves, at which time adoption might be more difficult. Only one or two pups per litter typically survive among wild dingoes; therefore, adoption may have increased both overall pup survival and the relationship between humans and wolves or dingoes. In the case of Australian Aboriginal people, there is evidence that human females would even suckle dingo pups (Meggitt 1965), which makes the issue of weaning moot. This suggests that while adult dingoes may not be better off staying with humans, pups might do better. Thus, the relationship may be more complex than simply "commensalism," with the dingoes being unaffected and the humans benefiting. This may represent a case of delayed return altruism (Rothstein and Pierotti 1988), with the pup paying back as an adult for benefits it received earlier in life.

One phenomenon we have discussed is that most domestic dogs (and breeds) retain wolf puppy characteristics their entire lives (Morey 1994, 2010). As a dog matures, it may experience some changes in behavior that are associated with maturity, but it rarely achieves the level of independence required by a wolf living in the wild. Still, it is important to keep in mind that wolves remain dependent upon their packs until they are at least two years old, often longer (Mech 1970).

While dingoes are young they often have very close relationships with humans but, like wolves, once they mature, their agendas are less centered on the interests of humans, although they remain friendly toward humans and do not fear them. This also seems to be true of camp dogs in Australia and wild dingoes, who when mature are less strongly attached to their human companions, even if they remain friendly. Formerly, Aboriginal "dogs" identified by Europeans in Australia were dingoes taken from their dens as pups and then raised by humans. Today, however, they are typically hybrids of many nondingo varieties. "Camp dogs are dependents . . . like children in that people give them skin identities, personal names, food, and shelter. Unlike children, they do not grow up to be responsible adults. In contrast, the wild dingo hunts his own food, makes his own camp, finds his own shelter, and follows his own law" (Rose 2000, 176).

The debate about differences between dingoes raised in camps and dingoes that lived in the wild has been challenged by some observers. For example, Lumholtz 1889 and Basedow 1925 (cited in Meggitt 1965, 17–18) agreed that the dingoes in camps were well fed. However, they also described dingoes in the camps as "pampered and relatively useless in hunting." According to Basedow, "A native just holds the unruly mob around him for company's sake; he prefers to rely on his own skill and instinct when hunting, and rarely allows his dogs to go with him; in fact, there seems to be little inclination on the part of the dogs to accompany the chase with the master." The picture painted by Basedow is of animals that appear to enjoy the company of the Aboriginals, just as the people enjoy having the dingoes around. Still, this description clearly represents a colonial attitude that mirrors the view that British invaders took toward the Aboriginal people themselves.

Aboriginals, in general, treat their dingoes very well (Meggitt 1965; Rose 2000, 2011). They incorporate warnings against killing dingoes within stories. The traditional story is simply a variation on "There was a man who shot dogs and he's dead now." The longer version of the story offers more detail. "There was a European stockman that had been mustering in an area that is important to Dingo life, and had ignored the advice Aboriginal people gave him. His shots were followed by a loud booming noise which signaled something out of the ordinary, and he later died" (Rose 2000, 29). This story bears considerable resemblance to the belief of the Koyukon people of northern Alaska regarding the provoking of a wolf spirit. "The wolf spirit is believed to be a powerful and potentially vengeful spirit. Disrespecting the wolf spirit can lead to bad luck, loss of children, and other tragedies" (Nelson 1983, 22, 158–62).

The relationship between dingoes and Aboriginals has evolved from being one completely centered on cooperative hunting to genuine companionship. It is common knowledge that dingoes usually return to the bush at some point; therefore, the major question is what keeps them around at all other than the bond they've formed with humans. These animals are fully capable of surviving on their own without the interference of humans.

Today many Aboriginal camps or townships in Australia are being overrun with domestic dogs. This is not a problem associated with dingoes, which partake of a natural cycle of returning to the wild. One reason they do so is that oddly enough, dingoes show a general inability to breed well in captivity (Corbett 1995). The issue is resolved by people allowing their dingoes to leave and breed out in the bush, leaving open the opportunity to return. Dingo pups born in the wild are often brought back to the camps to be raised. One wolflike character of dingoes is they are clearly biparental, and the male devotes considerable time and energy to feeding his mate and rearing his pups, a situation not observed among most contemporary domestic dogs (Corbett 1995: Spotte 2012; Coppinger and Feinstein 2015). At the same time, independence in breeding and mate choice, with no selective breeding, reduces the likelihood of the appearance of new traits and thus reduces the likelihood of changes in the morphology of the species.

In North America and Siberia, the relationship between humans and wolves has been going on even longer than the relationship between dingoes and Australian Aboriginal peoples, a minimum of 15,000 years (Morey 2014). The difference is that in these areas humans moved into areas already inhabited by wolves, and thus the "dogs" remained morphologically indistinguishable from wolves for long periods of time although they coexisted with humans in a similar sort of relationship. In contrast, in Australia humans had existed for long periods of time and the wolves/dingoes moved in on well-established human populations. In this case the lonesome humans were so pleased to have another placental companion, especially one who shared their social proclivities and hunting traditions, that they felt no need to change their new canine companions into a more domestic form, even if they were willing to raise and socialize puppies, therefore increasing overall breeding success in wild dingoes. After all, Australian Aboriginal peoples had no tradition of domesticating any kind of animal until they were forced into this way of life by European invaders.

CHAPTER SEVEN

North America

THE WORLD WOLF MADE

THE STORIES OF NORTH AMERICAN Indigenous peoples concerning wolves and their early experiences with this species can be divided into two basic categories (we don't discuss South America, because there are no true wolves, or even coyotes, on that continent): "those that occur in mythtime and those that occur in human time or historical time" (Bringhurst 2008, 169). This distinction is important.

In the first category are tribes that regard wolves as creator figures in their cultural traditions, which occur in "mythtime." The relationship these tribes have with wolves can be understood through close examination of many of their creation stories. For example, in the Shoshonean tradition (and language group), which includes Shoshones, Comanches, Paiutes, Utes, and Gosiutes, Wolf is considered the benign creator figure, and Coyote is her trickster younger brother (Wallace and Hoebel 1948; Ramsey 1977; Buller 1983, Pierotti 2011a, 2011b). Tribes that believe the wolf is a creator spirit have, understandably, a respectful relationship with wolves, as can be seen in many of the stories examined for this project. We look at these stories in greater detail in the next chapter dealing with creators and tricksters. Each story is dynamic in the sense that it is alive in the present moment. This means that the story is not a historic artifact but an expression of the contemporary perspective of the culture from which the story emerges.

Present in the second category of stories is the common theme of the wolf, either as a species or as individuals, being the savior or guide of the people. These stories seem to be located in historical time—they refer to events in the recent past or even in the present. Such stories often involve wolves acting as "teachers" to humans in hunting, allowing the people to first survive and then thrive within their new environment. Another type of story tells of a lost woman or child adopted by a pack of wolves before returning to home and people. What happens after the person returns is where much of the variation among tribes can be seen.

The creator and savior categories occasionally overlap. If the individual saved by wolves proves to be the last surviving member of his or her people, the wolves have ultimately saved all future generations of the tribe. One such story is "No Name," which comes from a Coastal Salish Nation on the shores of Tsla-a-wat, an inlet close by present-day Vancouver. The story begins at a time when the village was completely deserted and left in ruins:

> Everyone died—all except one tiny baby boy . . . too young even
> to have a name, so we will call him "No-name." . . . He would
> have died too if a she-wolf had not come by and, seeing no
> people about, started sniffing around the village. . . . No sounds
> of children playing or men busy at their work . . . only the
> calling of gulls and the grating "Krrkk, Krrkk" of two ravens
> wheeling high in the sky. . . . The she-wolf nosed around till she
> was satisfied that no harm threatened her. Then she trotted into
> the house where "No-name" was sleeping. She padded about,
> sniffing at the dried meat and fish hanging from the roof
> beams. She spent some time gnawing a deer bone which was
> lying in the cold ashes of a fireplace. Then, licking her lips, she
> wandered up to the baby's cradle and peered in [and] gave a little
> whine of surprise when she saw the boy. . . . When No-name
> woke up and began to whimper, she was reminded of her four
> cubs she had left in their den on the hillside. She gazed with
> steady yellow eyes at the tiny face, so hairless, which grew all
> puckered up as the whimper turned into a frightened wail. As
> she gazed, she sniffed. This must be a man-cub, she decided,
> recognizing the same scent she caught from hunters in the

forest. But where was his mother? The she-wolf hesitated. Surely this small thing must be cared for, and just as surely there was no one here to care for him. Quickly she made up her mind. She took the baby up in her powerful jaws and started for home. (Simeon 1977, 7–8)

The boy is raised by the family of wolves through early adulthood, when he eventually finds a human settlement and meets a young woman, whom he pairs with. The two move back to the village where the boy was born and start a family with the totem of a wolf to symbolize the respect the people have for the wolf who saved their lineage.

WOLVES AS TEACHERS

Many tribes that lived on the edges of the American forestland west of the Mississippi or on the Great Plains show a tradition in which wolves were regarded as role models and hunting teachers. These tribes include the Cheyenne, Lakota, Blackfoot, Assiniboine, Arikara, Arapaho, Pawnee, and especially the Shoshone (Hampton 1997).

The nature of such relationships is well illustrated by the Tsistsista (this is the true name of the Cheyenne, the term by which they refer to themselves [Schlesier 1987]), who appear to have had an ancient relationship with Wolf, dating back many thousands of years. According to a description from a Cheyenne elder, Cheyenne history was divided into four parts (Powell 1979), one of which was the "time of the dogs." (A point of clarification here: Europeans often mistranslated the term *wolf,* substituting *dog* when describing Indigenous cultures. Tribes often used the same term for both because their "dogs" were basically "wolves.") The four parts were the time before the tribe was struck by disease, followed by the time of the dogs (or wolves), the time of the buffalo, then finally the time of the horse. The horse and the buffalo were pivotal to the livelihood of the tribes living on the Great Plains, as is evident from stories and images recorded by both Native Americans and Europeans after contact between the two cultural traditions. Incorporation of the time of the dogs as a crucial historical period suggests that these animals were also of major importance in the early history of the Tsistsista. This period is less well known because the time of the dogs had come to an end before Europeans invaded and colonized North America. In order for the people to arrive at the time of the dogs, the relationship must have started much earlier with the ancient form of dogs, which means wolves. Dr. Henrietta Mann, a Cheyenne elder and pipe carrier, has stated that the "dogs" the tribe had at this time in their history were actually wolves (Pierotti 2011a).

A more detailed description comes from George Bent, the half-Cheyenne son of Colonel William Bent, who operated Bent's Fort, a trading post on the Upper Arkansas River in Colorado. George Bent kept the only written eyewitness account of mid-nineteenth-century interactions between Indians and Europeans from the tribal point of view. He was considered a reliable, accurate informant, and his information was verified by other accounts (G. E. Hyde 1968). Concerning the relationship between Tsistsistas and canids prior to the acquisition of horses, Bent states:

> The tribe had a great number of large dogs . . . employed to pack
> or drag burdens . . . used just as horses were in later times. . . .
> These dogs of the olden time were not like Indian dogs of today.
> They were just like wolves, they never barked, but howled. . . .
> Old people say that every morning just as day was breaking, the
> "dogs" of the camp, several hundred of them, would [gather] . . .
> and all howl together.
>
> Antelope Woman [a Tsistsista elder, described] . . . her
> mother [telling] her about the winter buffalo hunts . . . when
> all the tribe was on foot [i.e., before they had horses]. A herd
> of buffalo was surrounded by the people [and the "dogs";
> Schlesier 1987] and driven into deep drifts. . . . If a buffalo got
> away the dogs would set on it and quickly drive it back to the
> deep drifts. . . . After the buffalo are skinned [and butchered]
> the dogs [dragged] the bundles of meat over the ice. . . . As
> soon as the camp was reached, the dogs were loosed, and at
> once the whole pack rushed back . . . to the [kill site, where] . . .
> they feasted on the parts that had been thrown aside [during
> butchering]. . . . Mother dogs who had puppies in camp
> would run to the [site], gorge themselves with meat, and then
> run back to camp and disgorge part of the meat for the puppies
> to feed on. Sometimes a mother would make several trips to
> get enough meat for her litter of young ones. (G. E. Hyde
> 1968, 9–11)

FIGURE 7.1 Detail from *Four-Footed Allies*, by Cheyenne artist Merlin Little Thunder. (Courtesy of the artist)

This account reveals the wolflike nature of these "dogs" (figure 7.1). First, group howling initiated by the animals is a wolf trait. Although some dog breeds do howl, it is rarely a group activity. Cheyenne pack animals were large, strong beasts that howled at the approach of the morning like wolves (Powell 1979; Fogg, Howe, and Pierotti 2015). Second, the description of mixed-species cooperative hunting follows arguments made by Schlesier: "Cooperation with others [was taught] by the one animal that both the people of northern Siberia and the Tsistsistas regarded as the master hunter par excellence—the wolf" (1987, 35). Hunting ceremonies of both Tsistsistas and some Siberian peoples emphasize learning from wolves (Schlesier 1987). Finally, the description of parent animals returning to the kill site, filling themselves with meat and returning to regurgitate to pups is pure wolf. No dog is known to behave in this manner (Scott 1968; Spotte 2012; Coppinger and Feinstein 2015). This provides some sense of "the time of the dogs."

This issue of perception regarding "wolves" versus "dogs" warrants further consideration. Historian J. H. Elliot (1992) has argued that when Europeans first encountered the landscape, wildlife, and peoples of the Americas, they were so overwhelmed by novelty that they could not really perceive what was before their eyes. Instead, they insisted upon viewing American peoples and places through a European lens, seeing only what they expected to see, a situation we have described as Euro-bias. Thus when Europeans found people living with large canids they could only conceive of such animals as "dogs," and hence they described them as such and translated Indigenous terms for these animals as "dog." In contrast, the Indigenous people knew what type of creatures they were living with: "wolves." The fact that all scholarly work on this topic describes these animals as dogs might also be attributed to a phenomenon attested to by Thomas Kuhn, philosopher and historian of science: "Closely examined, whether historically or in the contemporary laboratory [scientific research] seems an attempt to force nature into the preformed and relatively inflexible box that the paradigm supplies. No part of the aim of normal science is to call forth new sorts of phenomena; indeed, those that will not fit into the box are often not seen at all" (1996, 163) Perhaps explorers of European ancestry could not accept the evidence before their eyes—that Indigenous Americans were living with socialized wolves. One European visiting tribes stated: "The Indian dogs which I saw here so closely resemble wild wolves that I feel assured that if I was to meet with one of them in the woods I should most assuredly kill it as such" (Audubon 1960, 520). Early European visitors decided that because these animals lived with humans, they must fit into the European concept of domesticated *Canis*—dogs. Contemporary scholars perpetuate such errors because they assume that only written statements by Europeans count as true evidence (Pierotti 2011a). Despite claims to impartiality, most scholars cannot see beyond the bonds imposed by their narrow cultural perceptions (Ritvo 2010).

To return to our theme, the relationship between the Cheyenne people and wolves persisted until recent times. After the Sand Creek Massacre in November 1864 carried out by the "Reverend" John Chivington and his followers, two Cheyenne women and their children managed to escape. The group took refuge in a shallow cave under a bluff. In the night, a large male wolf entered the cave and lay down beside them. In the morning the wolf traveled with them, stopping to rest whenever they did. When one woman addressed the wolf, telling him of their need for food, the wolf led the group to a freshly killed buffalo. Over the next few weeks, the wolf stayed at their side, catching food and protecting them from potential enemies both human and nonhuman. Finally, the wolf led them to a Cheyenne camp on the Republican River. After delivering the women and children to safety and being fed in return for his services, the wolf left (Grinnell 1926, 149–53).

Some Tsistsistas could "understand the speech" of wolves. By listening to wolves howl, they could anticipate events to come and warn or prepare their fellows. They were said to have gained these skills from living with the wolves, and if ever they were lost, hungry, or near death the wolves would rescue them. This is similar to the Lakota story "The Woman Who Lived with Wolves" (Marshall 1995). Given the way Indigenous peoples were being treated by their fellow humans, these stories, which are quite common among Plains people, reveal that nonhumans can be better friends and companions than some humans.

The Blackfoot people (NiiTsitapiiksi) also credited wolves with having taught them to be hunters, especially in regard to buffalo drives. The wolves would cut a single bison out from the herd and either run it until it became exhausted or drive it up a steep slope and harass it until it either succumbed or jumped to its death, at which time the wolves would circle down and feed upon it. The Blackfoot duplicated this behavior in their own drives, down to small details. Another old story from the Blackfoot people describes how a family on the verge of starvation was discovered by a pack of wolves, which provided the humans with food until they were able to hunt on their own (Hernandez 2013).

Blackfoot were fond of wolves as companions. In preparation for hunts they slept on wolf skins and sang songs to encourage wolves to join them. If a wolf howled nearby while a hunting party was on the move, hunters would sing, "No I will not give you my body to eat, but I will give you the body of someone else, if you will join us" (Grinnell 1892, 260–61). According to a traditional Blackfoot story:

> Once there was a SiksikaiTsistapi [who] . . . had but one horse. He was a poor man. He was very fortunate . . . on this day he had a kill. He cut up the meet and packed it on his horse. Also, he left enough for a good meal for . . . the wolves. . . . He ran into a pack of wolves . . . on their way to the remains of his kill. . . . Later, he ran into an Old Wolf who was having a hard time keeping up with the pack. The hunter stopped and offered the choicest cut to the Old Man and told him, "By the time you get there, there may be nothing left for you to eat." . . . It was a

cold winter and everyone was hungry, [the Old Wolf] replied, "I am in a hurry, those ahead are hungry. I need to get there because they will not start without me. You see, I am the Grandfather. You will receive a gift for your generosity." Later the hunter was very fortunate in his hunting. (Bastien 2004, 35–36)

Like the Cheyenne, the Blackfeet identify a period called the Era of the Dog (*Iitotasimahpi Iimitaiks*), also called the time of the ancestors, which refers to the period before contact with Europeans. This, preceding the time of the horse, was a period when they depended upon their dogs (wolves). These animals were given great respect because they were companions of humans possessing spirit and consciousness (Bastien 2004, 8–14).

Another traditional story from the Blackfoot people is "The Legend of the Friendly Medicine Wolf":

A large party of Crow Indians . . . attacked [and] . . . the Crows . . . carried away some women prisoners. One . . . was a young woman named *Itsa-pich-kaupe* (Sits-by-the-door). . . . She was carried on horseback . . . over two hundred miles to the Crow camp. . . . *Itsa-pich-kaupe* was so closely watched she could find no chance of escape. . . . One day . . . the two women were together. . . . The Crow woman conversed with *Itsa-pich-kaupe*. . . . "I overheard my husband say they are planning to kill you. I . . . will help you to escape tonight when it is dark." . . . When . . . the Crow woman knew . . . that her husband was asleep, she crawled over to *Itsa-pich-kaupe* and unfastened the ropes . . . [and] tied the loose end of the waist-rope to a lodge pole . . . loosened the bottom of the lodge covering . . . giving *Itsa-pich-kaupe* a pair of moccasins, a flint and small sack filled with pemmican. . . . When she . . . was a long distance from Crow camp, she began travelling by day. . . . She saw a large wolf following. . . . The wolf stood watching her. . . . When *Itsa-pich-kaupe* arose . . . the wolf followed. . . . She sat down again to rest, he lay down by her side. . . . "Pity me brother wolf! I am so weak for food that I must soon die. I pray for the sake of my young children that you will help me." . . . The wolf . . . soon came back, dragging a buffalo calf. . . . After roasting and eating some of the meat, she felt stronger. . . . The wolf drew near, she placed her hand on his broad back, and he seemed glad to bear her weight. . . . He brought her safely home. When they entered camp together . . . she related . . . the story of her escape. . . . She besought the people to be kind to the wolf, and to give him some food. The faithful wolf waited for a long time watching in vain for *Itsa-pich-kaupe*. . . . Her relatives continued to feed him until he disappeared, never to return. The Blackfeet never shoot at a wolf, or coyote, believing them to be good medicine. (McClintock 1910, 473–76)

Both these stories indicate solid cooperation and a form of reciprocity between members of the tribe and wolves, linked to guiding and the sharing of food. The latter story recalls the Cheyenne narrative about the aftermath of Sand Creek. Another story from Brings Down the Sun describes his reaction to hearing a wolf howl: "We consider the wolf a friend of man, and do not believe it is right to shoot him. We have a saying, 'the gun that fires upon a wolf or coyote will never again shoot straight.' Did you ever know of a wolf who did not wander? They never stay long in one locality. They raise their young in one place and then go to another. They are continually roving over the country and are always on the move. My father named me Running Wolf, and . . . I am like the wolf, for I love to roam over the prairies and among the mountains" (McClintock 1910, 434).

One point should be obvious from all these stories: when Wolf performs service and guides or provides for people, it is always fed in return. This point is important because it is has been argued by the Coppingers and their adherents that the first wolves associated with humans scavenged or hung around camps waiting for scraps (e.g., Coppinger and Coppinger 2001). It is quite obvious, at least from the stories told by Indigenous American peoples, that humans voluntarily fed or shared food with the wolves with whom they associated. Such traditions go back thousands of years. As a well-known folklorist of Native American cultural traditions states, "My experience is limited to the narrow scope of my lifetime within a culture that is practiced in a continuity of decades, generations and lifetimes. [In contrast] the traditions of these [Indigenous] societies are a process measured in centuries" (Welsch

1992, 27). This difference in experience and worldviews creates a problem. Euro-Americans see the world in small timescales and therefore, thinking in these terms, to them the origin of dog breeds in the last couple of centuries can be considered equivalent to the beginnings of domestication, when in fact in that time frame we are looking only at the most recent stage of an ancient tradition.

The Lakota (Sioux) also tell a story of an injured woman who is saved by wolves (Marshall 1995). Once the woman has healed she returns to her people with a valuable set of skills taught by the wolves. Lakota scholars and writers, such as Joseph Marshall III (1995), claim the Lakota have a brotherly relationship with the wolf because they learned their hunting skills from the wolf. This practice of learning and being instructed by wolves is demonstrated in the story by the many instructions the leader of the wolf pack gives the young woman regarding the return to her people. The wolves provide everything the woman needs. At no point in the story does the woman take a leadership role in the pack, which is always assumed to be an essential component of the domestication of wolves as they changed to dogs.

One of the biggest problems with determining when domestication actually took place is that for much of history, dogs kept by Indigenous groups were impossible to differentiate from wolves through their physical remains. "Early animals called dogs are . . . phantoms. . . . There were no dogs at that time. . . . Although there may have been socialized wolves, they were not dogs. The lack of physical evidence is due to many reasons, not the least of which is that the dog existed genetically before it did phenotypically . . . because the dog is fundamentally a wolf still capable of looking like a wolf" (Derr 2011, 37). One explorer of the upper Missouri River in 1811 wrote, "The Arikara dog is nothing more than the domesticated wolf. In wandering the prairies, I have often mistaken wolves for Indian dogs" (Brackenridge 1904, 115). Prince Maximilian, traveling in the western United States in the nineteenth century, reported: "In shape [the dogs] differ very little from the wolf, and are equally large and strong. . . . Their voice is . . . but a howl, like that of the wolf" (1906, 310). Such reports reinforce our argument that a primary criterion many European visitors used for distinguishing wolves from dogs was whether the canid in question was associating with humans or not. If it was, it was a dog; if not, it was a wolf (Fogg, Howe, and Pierotti 2015).

Ronald Nowak argues that gray wolves did not live in the central areas of what is now the United States until around 10,000 YBP (cited in Beeland 2013), which would mean that they arrived in this particular habitat around the same time as did the ancestors of many Plains tribes. The interesting implication is that humans and wolves may have entered many parts of the Americas as cooperating species working both together and independently to learn new ecological situations. In the stories we recount, the humans are still in the role of the pupil while the wolf is the instructor.

The Tsistsista believe their tribe was taught to hunt by two sacred wolves. The first, identified as male, is "the wolf *maiyun,* the species-specific protector spirit of wolves and his female companion is the Horned Wolf." The two were "the master hunters of the grasslands and, with their species, the protectors of all animals." Maiyun chose to teach the newcomers, the humans, to hunt on the grasslands. "As the 'invitation song' of wolves called raven, coyote, and fox to share in their kill, so did Tsistsista hunters call wolves to their kill or set meat aside for their use" (Schlesier 1987, 82).

Early humans may have learned their hunting techniques from mimicking the methods used by wolf packs. This could involve simply wearing the skins of wolves in order to get closer to the prey species without alarming them; the appearance of a lone wolf would not startle prey as much as seeing a human. This practice has been discussed explicitly by Lakota scholar Joseph Marshall III: "Many Plains hunters who hunted the buffalo often crawled on their hands and knees, singly or in pairs, to the very edge of a herd, many times wearing wolf capes, and mimicking the movements and mannerisms of wolves. Since Buffalo had little to fear from one or two wolves, they often watched in curiosity, thus allowing hunters to get well within shooting range" (1995, 15). The activity is illustrated in a painting by George Catlin known as *Buffalo Hunt, Under the White Wolf Skin, 1844.* (Pierotti has in his personal collection a version of this work by Comanche artist Blackbear Boisin.) In this painting two hunters are draped in the skins of two white wolves. Catlin (1973) describes the technique from his firsthand observation: "While the herd of buffaloe are together, they seem to have little dread of the wolf, and allow them to come in close company with them. The Indian then has taken advantage of this fact, and often places himself under the skin of this animal, and crawls for half a mile or more on his hands and knees, until he approaches within a few rods of the unsuspecting group, and easily shoots down the fattest of the throng."

Some ramifications of this tradition are illustrated by a story about a Dakota hunter Joseph Marshall III tells in *On Behalf of the Wolf and the First Peoples:*

> The hunter waited in ambush and shot a buffalo with several arrows. Of course the buffalo did not immediately die, so the hunter had to follow the wounded animal. The buffalo finally collapsed, and as the hunter hid and waited a safe distance to make sure the animal had expired, a wolf appeared and warily approached the buffalo. Displaying the utmost patience and caution, the wolf moved only a step at a time. Finally, she reached the downed buffalo, which by then had died. The wolf's demeanor and posture told the hidden hunter that the buffalo was dead and therefore, safe to approach. But out of curiosity he waited to see what the wolf would do, fully expecting her to begin tearing at the flesh with her fangs. Instead, she went around and around the carcass until she saw the arrows protruding from the buffalo's side. She sniffed the arrows then sat back on her haunches to carefully test the wind. After a time, she looked directly at the hunter's hiding place with a long penetrating stare, and then nonchalantly walked away from the dead buffalo and disappeared over a rise. Later, after his wife and family had butchered the buffalo, the hunter made sure that they left behind some choice portions to share with the wolf and her family. (1995, 12)

Such stories demonstrate that among at least three major peoples of the American Great Plains reciprocal relationships between humans and wolves were demonstrated through the sharing of spoils from successful hunts. The protector spirit of a species, mentioned earlier, is considered an individual who has the power to punish hunters for abusing animals under his protection by withholding game or by inflicting injury on the offender (Schlesier 1987, 4), so this offering served as a way to show respect. Such entities were common concepts among Indigenous peoples of North America (Pierotti 2010).

The Tonkawa people of east central Texas also had good relationships with wolves. Tonkawa means "the people of the Wolf," and they claimed they were all descended from a mythical wolf. The Tonkawa would never kill a wolf, and they refused to farm because they were wolves and wolves hunt for their food (Moore 2015).

Another Plains people who personified and emulated wolves were the Skidi, the largest band of the Pawnee, who lived along the Loup (Wolf) River in present-day Nebraska. Skidi were respected for their ability to travel all day and night, living by scavenging carcasses or not eating if no carcasses were available. In response the Plains tribal sign language sign for wolf, basically a contemporary peace sign held beside the right shoulder and moved forward and upward, also became the sign for the Pawnee people themselves. The Pawnee also called their war parties *araris taka,* or "white wolf gangs" (Hampton 1997).

In recent conservation efforts, the Nez Perce tribe in Idaho has been stalwart in its support for the reintroduction of the Yellowstone wolf, and the White Mountain Apache in Arizona have allowed the reintroduction of wolves onto their tribal lands (Pavlik 1999; Ohlson 2005). The wolf is an integral part of the history and culture of the Nimiipuu, Nez Perce for "the people," and the return of the wolf is manifestation of the tribe's spirituality, culture, and history (Ohlson 2005).

Such relationships are not restricted to the Plains. Wolves are known by the Koyukon people of central Alaska for intelligence, strength, keen senses, and hunting skills (Nelson 1983). The Koyukon also believe that they learned their hunting skills long ago from wolves. The elders pass on to younger members of their society rules concerning the proper treatment of wolves: eating, skinning, burning, or any other use of a wolf carcass is banned. "For the Koyukon, the similarity between wolves and humans is no coincidence—in the distant time, a wolf person lived among people and hunted with them. When they parted ways, they agreed that wolves would sometimes make kills for people or drive game to them, as a repayment given when wolves were still human" (159). Koyukon people believe the name of the wolf is not to be said out loud because it summons the spirit, which is powerful and potentially vengeful. Disrespecting the wolf spirit can lead to bad luck, loss of children, and other tragedies. This did not mean, however, that given the agreement alluded to above, the Koyukon would not take wolf kills if they encountered them unattended.

Among the Ojibwe of northern forests, connections with wolves appear in early stories:

> When Original Man first walks the earth, he complains to the Creator that all the animals are paired. Yet he is alone. The Creator gives him a companion, Ma-en-gun (the wolf). After roaming the world together and seeing all things under creation, the creator says they must separate and adds: "What shall happen to one of you shall happen to the other. Each of you will be feared, respected, and misunderstood by the people that will later join you. . . ." This last teaching about the wolf is important to us today. . . . Both the Indian and the wolf have come to be alike and have experienced the same thing. . . . Both have a clan system and a tribe. Both have been hunted for their wee-nes-se-see' (Hair). And both have been pushed very close to destruction. (Benton-Benai 1979, 8)

In the southeastern United States, Lawson (1967) reports that the Waxhaw tribe of North Carolina kept animals described as "wolves." This shows a person of European ancestry acknowledging that the animals living with the tribe were "wolves." It is likely that, given the location, the story refers to the red wolf, *Canis rufus,* rather than to *Canis lupus.* Red wolves, as Pierotti has observed, are not as compatible with humans as gray wolves because they are less easily socialized (Lopez 1978).

In the Pacific Northwest, Wolf is considered to be the founder of many lineages among First Nations peoples. The Tlingit divide themselves into two "modalities," Wolf and Raven, which probably reflects the relationship between these two species (as discussed in our chapter 2). Raven is considered the creator figure by these peoples, but Wolf is still an important companion and role model. For example, the Tlingit tell a story of fishermen who come across an exhausted wolf swimming far from shore. They take it into their boat, and it becomes their constant companion, hunting and living with them until its death (Garfield and Forrest 1961, 105). The Tlingit also learned how to stalk seals from wolves—the wolves take a mouthful of seaweed to break up their swimming outline and swim up to seals with their heads thus disguised (Muir 1916, 138).

The Haida, the most seagoing Northwest First Nations people, tell a story about hunting different from that of most other tribes. In their story,

"The Waskos: Dogs of the Sea," the wolf is a sea wolf and actually feeds the tribe by killing whales and dragging them onshore: "One day, when the Indian people of Hunter's Point were short of food, the two Waskos ran into the sea and started to swim, using their strong front legs as paddles. Sadly, brother and sister watched their pets swim right out of sight. 'They'll be drowned,' sobbed Sister. 'They'll never come back,' cried Brother. Father . . . had a feeling that the Waskos would bring good to the tribe. The Waskos did return a few hours later, each carrying three whales. Every morning the great dogs swam out to sea, bringing so many whales that there was hardly enough room on the beach. Everyone in the village was busy from morning to night, cutting up the whales, feasting, and storing the good food for the winter" (Simeon 1977, 14, 16). Again, the same themes of kindness and loyalty are present in the story, with the variation coming only from the setting because the legend comes from a coastal people. This story may refer to orcas (*Orcinus orca*), socially hunting cetaceans that have been described as "wolves of the sea," and are also known to have cooperative hunting relationships with humans (Clode 2002; Pierotti 2011a).

Among the Inuit people, shamans would sing songs learned from wolves because the wolf was considered to be the most important communicator with the *tunraq*, or spirit of the universe. To share abilities with wolves was a major way of communing with this unknown power (Ray 1967).

SHEEPEATERS (THE LAST FREE SHOSHONI)

One of the best documented examples of the relationship of a people with wolf/dogs comes from the T'ukudeka Newana, or Sheepeater (Mountain) Sosoni (Shoshoni), who lived in the higher mountains of the Intermountain West, including the area today known as Yellowstone, into the current state of Idaho along the Continental Divide (Corless 1990; Loendorf and Stone 2006). One group, the so-called Weiser Shoshoni, lived independent of white government in the mountains of western Idaho between the Salmon and Payette Rivers until almost the beginning of the twentieth century (Corless 1990).

Sheepeater dogs, large and robust animals who worked alongside their human companions, were integral components of band society. "Their coloring revealed that there were wolves in their family tree, not only long ago but also more recently" (Loendorf and Stone 2006, 103). The resemblance of these dogs to wolves was described in numerous historical accounts, including that of John Richardson, who in 1836 wrote that "wolves and the domestic dogs of the fur countries are so much like each other that it is not easy to distinguish them at a small distance" (cited in Loendorf and Stone 2006, 104). Friedrich Kurz, who traveled in the Yellowstone region in 1851, noted that these dogs "differ very slightly from wolves, howl like them, do not bark, and not infrequently mate with them" (1937, 239). The primary difference between these animals and wild wolves is that they were stronger in the shoulders, which was necessary for their ability to haul travois loads of seventy pounds and to carry packs weighing as much as fifty pounds.

Indians on the northern Plains intentionally made their dogs accessible to wolves for breeding (Loendorf and Stone 2006). The Shoshone depended upon their dogs as beasts of burden; therefore, Sheepeater hunters periodically introduced wolf genes into the bloodline of their four-legged partners to improve their strength and endurance. Examination of dog skeletons found in Yellowstone Park established that the dogs ranged in height from medium to tall, placing them between coyotes and wolves on a scale of stature (Haag 1956). Their frame was robust and supported a large head that was comparable to that of a wolf (figure 7.2).

The quality of life and behavior of Sheepeater dogs was described by Osborne Russell (1914), who recorded his encounter with a small group of Sheepeaters in Yellowstone National Park in 1835: six men, seven women, and eight to ten children, accompanied by thirty dogs. This ratio of more than two dogs to every adult represents many mouths to feed. Russell noted that the dogs appeared to be well fed, well behaved, and contented, and that Sheepeaters customarily fed their dogs before they themselves ate. This reflects a theme discussed above, that Indigenous people made an effort to see that wolves they encountered and their wolflike dogs were treated as equals. This suggests a tradition that may have been thousands of years old and explains the benign relationships between humans and wolves throughout both Asia and North America. The Japanese, especially the Ainu, also had a tradition of feeding wolves or wolflike dogs as a means of maintaining a cordial relationship (chapter 5; Walker 2005).

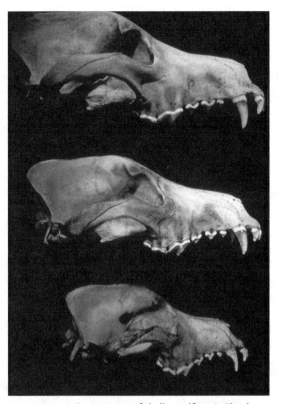

FIGURE 7.2 Comparison of skulls: wolf (*top*), Shoshone
"dog" from Yellowstone dig (*middle*), domestic dog
(*bottom*). Note the overall similarity of the top and middle
skulls, which have similar proportions, pronounced
sagittal crests, enlarged craniums, and bigger teeth,
especially the canines. (Yellowstone Photo Collection)

Sheepeater dogs were day and night guardians. Imagine the uproar
that thirty dogs would have made when aroused. Their vocalizations
served a protective purpose, warning of the approach of another tribe's
war party or driving off wild animals such as grizzly bears and mountain
lions (Loendorf and Stone 2006).

Trappers and travelers who visited Crow, Flathead, Nez Perce, and
other western Indian groups in the 1800s all report that the use of dog
travois survived into the historic era even among horse-riding tribes
(Loendorf and Stone 2006). For families not affluent enough to own and
maintain horses, dog transport represented a cheaper alternative, especially for lighter loads such as carrying wood near camp. Dogs were more

adept than horses at movement in heavy winter snow, and could add their body warmth to the family bed on those nights when temperatures were below zero.

The packs used for dogs consisted of rawhide envelopes, or parfleches, lashed directly to a dog's back with rawhide binding that was strapped across the chest and under the belly (Shimkin 1986). Another strap went under the dog's tail to help secure the pack (see figure 7.1). Recent interviews with descendants of Sheepeaters revealed that they even made rawhide booties for their dogs to wear in the winter to protect their feet from frostbite and prevent ice buildup between the toes (Loendorf and Stone 2006).

The partnership between Sheepeaters and their dogs was most apparent when it was time to provision the family. Sheepeater men and dogs primarily pursued bighorn sheep and deer, working collaboratively to herd and then capture their prey (Loendorf and Stone 2006). The nature of the relationship between humans and their dogs among groups in the Plains and Rocky Mountain regions is demonstrated by the fact that Sheepeater dogs were well rewarded for their labor with generous portions of the meat they helped their owners obtain. The respect that Sheepeaters had for their dogs' intelligence and prowess is exemplified by the fact that two of the canine skeletons from Yellowstone Park described earlier were buried with an adult man, and another accompanied the burial of an adult woman (Loendorf and Stone 2006). That both men and women were interred with dogs suggests that both genders valued the contribution dogs made to all aspects of life and thought of their company as essential to the afterlife (see Morey 2010). Because we have historical records of the interaction, we know that this strong bond was forged early as Sheepeater children played with pups and raised them as friends whom they loved throughout their lives.

The Comanche, or Nuhmuhnuh, who are closely related to the Shoshone, lived on the high plains of present-day Colorado, Kansas, Oklahoma, New Mexico, and Texas. Nuhmuhnuh believe that "practically all animals and insects were attributed powers by someone" (Wallace and Hoebel 1948). The only two animals that were not thought to be capable of providing any powers were the horse and the dog. (Nuhmuhnuh distinguished between wolves and dogs in this regard: "the Comanche believe Wolf could grant the power to walk barefoot on the cold snow" [203]). The reason horses and dogs could not provide power was that "they were everyday parts of the Comanche household" over which humans had control, and "givers of power had to be outside [humans'] realm of dominance" (204).

The Nuhmuhnuh and Newana were the same people until after Europeans arrived, bringing both horses and domestic dogs. After a quarrel among leaders, the Nuhmuhnuh separated from the Newana, moving onto the Plains in the seventeenth century and evolving into a horse culture. As a result, Nuhmuhnuh did not have a tradition in which dogs and horses were not important parts of their culture, and they emphasized horses over dogs (Wallace and Hoebel 1948). The Nuhmuhnuh also lost the Newana transformer stories about the importance of Wolf and Coyote; "all that remained of the Coyote folk tale and myth cycle was that part that told of Coyote as a harmless bungler" (194). This difference between closely related peoples who shared a common language provides some insight into the way various tribes saw the relationship between wolves and dogs. A connection between the two was recognized by Nuhmuhnuh, who were unusual among Plains peoples for their adamant refusal to use dogs as food, even if they were starving, because one does not eat the creator's brother (69), although they readily ate horses. The wolf could provide powers to humans should they incorporate some quality of the animal, but the dog, even though recognized as a close relative of the wolf, lacked this ability.

As representatives of their creator figure (see chapter 8), wolves were not beings over which Nuhmuhnuh people considered themselves dominant. Humans lived and worked closely with dog/wolves in a relationship that probably represents the beginnings of one place of domestication. Wolves were not tame pets that required the dominance of humans like dogs introduced by Europeans. As discussed above, domestication should not be equated with loss of independence or the ability to live separate from humans. In the early stages of domestication many groups around the world had close relationships with wolves that were maintained throughout (pre)history. Only when wolves functionally became dogs in the last 10,000 years or so did humans come to see themselves as the dominant species in the relationship. This difference lies at the core of the debate among archaeologists discussed in chapter 4. More important, it reveals that living with wolves and the process of domestication were still taking place in North America prior to European contact within the last 500 years or so, showing that the assumption that dog domestication ended more than 10,000 YBP is overly simplistic.

HUMANS AS WOLF'S STUDENTS

Stories told by Indigenous American peoples about their relationships with wolves suggest that through much of the early stages of the evolution of human hunting practices, the wolf was the teacher and may have taken the lead in initiating many hunts, and thus they were not being directly manipulated by humans. Humans served as pupils in need of instruction, which casts a different light on the idea of how domestication may have proceeded, in that until recently humans were at best partners, even students, of wolves.

Marshall (1995) gives a specific account of this relationship, stating that the reason for the close relationship between wolves and the Lakota is the parallel aspects of the lives the two species share. "The people of that time learned the ways of the wolf because they understood the reality of their existence. Among them it was the hunter and the warrior who followed most closely the path of the wolf. As a hunter the wolf had no equal—with his sharp sense of smell, keen eyesight, and powerful jaws. Those were formidable weapons, but the first peoples saw that they were of little use without endurance, patience, and perseverance. They were even more important weapons of the wolf, and they were qualities the first peoples could develop in themselves" (6). As discussed in chapter 3, the family structure of a pack of wolves is very similar to that of a human nuclear family. These similarities are discussed in another of Marshall's works: "The [wolf] family was led by a bloka, or 'male,' and a winyela or 'female,' labeled the 'alpha' by non-native observers. The bloka and winyela had a litter of young usually every year. Those young stayed after they were weaned and grew to young adulthood. So the core family was several generations of offspring, but only the bloka and winyela mated and bore young. Usually, as the offspring grew into adulthood, they went off to form their own families" (2005, 35).

As Mark Derr points out, "The fault line created in the Anglo-European world between wild and domestic . . . underlie[s] many of our assumptions and attitudes, despite having been well exposed and discounted in recent years. Until all preconceived notions are laid aside . . .

we will not gain a clear understanding of the nature of the animal [dog] who fills so many different, frequently contradictory roles in quite different human societies" (2011, 85–86). The consistent pattern within all of these Indigenous stories is that the relationship between wolf and human was based upon respect and cooperation, especially in hunting, implying a complex and interesting alliance with the organisms that we refer to today as dogs. Indigenous understanding is illustrated by one Seneca storyteller describing the relationship between humans and dogs in the following way: "It is true that whenever a person loves a dog he derives great power from it. Dogs still know all we say; only they are not at liberty to speak. If you do not love a dog, he has the power to injure you with his *orenda* [magic]" (Schwartz 1997, 22).

It is clear that although attitudes concerning wolves vary among tribes to an extent, many of the same themes or viewpoints are present in most of their stories. These stories credit the wolf with being a figure from which the human world can learn. Companionship resulted from this amicable relationship; however, the most important contribution wolves and dogs made to the lives of Indigenous peoples in North America was their assistance as hunters, guardians, and beasts of burden.

THE ROLE OF COMPANION WOLVES AMONG
INDIGENOUS AMERICANS

We have discussed several reasons why wolves appear to be "pre-adapted to domestication" or, more appropriately, "predisposed to live closely with humans, which is the precursor to domestication. They live naturally in complex social groups. They are intelligent, highly adaptable, loyal, form pair bonds and hunt cooperatively in packs" (Gade 2002). All of these reasons eventually led to a beneficial situation whose rewards could be reaped by Indigenous American peoples.

"The first dogs probably helped human hunters find or track prey and protect their home territory and their social group (which included humans), behaviors that are common among living canids such as wolves, jackals, or coyotes" (Shipman 2011, 209). The most useful beneficial use of wolves/dogs varied depending on the lifestyle of the people involved. Westward migration of the Cheyenne/Tsistsista and Sioux/ Lakota was made possible by the use of dogs as beasts of burden before horses became available as a result of European introductions in the seventeenth century (Hassrick 1964, 157). These "travois dogs" allowed the Lakota and Tsistsistas to follow the buffalo herds across the Plains. "The hunting cultures generally used dogs for hauling and transportation" (Gade 2002). Following a moving herd across the Plains on hunting expeditions required frequently packing and moving equipment, such as tipis, clothing, and cooking implements. Greater distances could be covered with the use of travois dogs compared to the use of human power alone. According to one Lakota elder, "Travois dogs were trained to sleep at the door of the tipi and bark at strange noises. A simple 'tsst' [also the sound used by "Dog Whisperer" Cesar Millan to quiet unruly dogs; see Millan and Peltier 2006] was enough to stop the barking, but you had to hit them with a stick to stop them from annoying visitors" (Hassrick 1964,158). One factor evident in this passage is the role of protection undertaken by travois dogs. Dogs clearly had also been absorbed into the human culture enough to be invited into the living quarters of the humans. This is not a common practice with wild animals, and it shows that these humans were past the point of allowing the wolves/dogs to return to the wild.

Travois dogs were the first group or breed that was obviously domesticated among Indigenous Americans—for example, those with the Sheepeater Shoshone that made following the buffalo herds possible. The role dogs played in the lives of the Lakota is depicted in the account of Little Day, who described the doghouse made for a small female behind the tipi. These shelters were very similar in structure to a sweat lodge and were typically constructed by women. According to Little Day, "Nearly every family had one made in the same way" (Hassrick 1964, 158).

According to one story from the Cheyenne, dogs used as pack animals were "large, strong beasts that howled at the approach of morning like their relatives the wolves" (Powell 1979, 20). This shows the people made some distinction between their dogs and wolves that lived away from people. This split appears to have been a recent separation, if many dogs still retained some specifically wolf behaviors such as group howling. It should always be kept in mind that Europeans often refer to animals that live with people as dogs rather than wolves, even when the people themselves do not make such a distinction. For example, the Oglala Sioux formed a warrior group they referred to as the Wolf Society, but whites (Europeans) mistranslated this name as "Dog Society" (Hampton 1997,

48), and a similar error was made in naming the "Dog Soldiers" of the Cheyenne which, given the cultural traditions described above, were clearly "Wolf Soldiers."

We have discussed the traditions of a wide range of tribes that tell how they learned to hunt from wolves or by hunting alongside them, but this describes a more ancient relationship. In more modern times the hunting was obviously done cooperatively. Wolves were able to help in ways that other humans could not. For example, the sense of smell is much stronger in canids than in humans, so hunters who solicited the help of dogs were able to track game over much larger areas. Their sense of smell allowed hunters to hunt game that was not easily visible. It is much easier for an animal to hide from a hunter who is relying solely on one sense, eyesight, than it is to evade a hunter with a powerful nose combined with good hearing and keen eyesight. This increased aware-ness of the surrounding habitat provided the hunter with a certain amount of protection as well. Dogs also could grab and distract prey while humans made the kill. Finally and most important, dogs could run down prey that humans could not.

A similar theme is found in the image by the Cheyenne artist Merlin Little Thunder, *Eyes of a Sleeping Village* (see frontispiece), which illustrates the common goal of wolves and humans to protect the village at night. Another point illustrated well by this picture is the equality between the species. If this were a picture of a human and dogs, we would expect the human to be in the top position, as if looking over the dogs. In this image one of the wolves is at the top of the picture with the human below. The human is dressed as a wolf, and the wolves in the image are arranged in somewhat of a circle. All of these aspects represent how humans saw them-selves within this relationship. The humans had to learn from the wolves, which is why the human is imitating the wolf. The wolves could guard the village much better than the humans, but the human in the picture is working cooperatively with the wolves. The man dresses as one of the wolves to symbolize the respect he has for the abilities of the wolf. This theme is similar to that seen in the Catlin painting *Under the White Wolf Skin:* both images show humans attempting to do something wolves do better by imitating their canid companions.

CHAPTER EIGHT

Wolves and Coyotes

CREATORS AND TRICKSTERS

IN PREVIOUS CHAPTERS we concentrated on the relationship between *Canis lupus* and *Homo sapiens*, but in North America a new character enters the dynamic that appears to have significant influence on how our two major protagonists interact. This is the coyote, *Canis latrans*, a close relative of *Canis lupus* but very different in social structure and ecology, both of which render it an unlikely, if not impossible, partner in domestication.

Certain animals have characteristics that make them better candidates for domestication. Most of these relate to making animals suitable as livestock, as with domestic ungulates and the galliform birds (chickens, turkeys, guinea fowl) (Hemmer 1990; Ritvo 2010). In a few cases, we have succeeded in domesticating intelligent social creatures like elephants and horses. The only time we have successfully domesticated a social carnivore, which are relatively uncommon, is with *Canis lupus*. Although coyotes resemble wolves in general appearance, they are not truly social and do not create the kind of group structure that allows wolves to readily adjust to sharing their lives with humans. Coyotes are highly intelligent and supreme survivors; they know how to live away from humans as well as in the interstices of our society and ecology without trusting or relying upon humans (Dobie 1961; Fox 1971; Bright 1993; Papanikolas 1995).

Coyotes drew the attention of Native American peoples, although it is clear that they regarded coyotes as very different entities than wolves. They expressed this difference between the two species frequently in their stories, in thousands of accounts from hundreds of different peoples (e.g., Ramsey 1977; Ude 1981; Buller 1983; Haile and Luckert 1984; Mourning Dove and Guie 1990; Bright 1993; Berk and Anderson 2008). Coyotes are always regarded as complex and intelligent but antisocial. One theme that emerges is that they are regularly stirring up trouble by "trying to make things better."

At times, there seems to be confusion, or even conflation, when Euro-Americans try to distinguish between wolves and coyotes because the coyote looks like, and in many ways is, a small form of wolf (D. E. Wilson and Reeder 1993). Ronald Nowak, an expert on the systematics of American canids, has said that if you lined all the wolf and coyote skulls in museum collections in a continuous series, there would be no apparent break (personal communication). In eastern North America, "coyotes" grow to be almost as large as small wolves; therefore, using size as a distinguishing characteristic can be unreliable. The eastern coyote, a new kind of canid that remained unidentified until the 1990s, is between gray wolves and coyotes in size and seems to be a complex of several varieties of *Canis* that extend from the Appalachians through the Adirondacks up into the Canadian Shield (P. J. Wilson et al. 2000, 2001, 2009).

Indigenous peoples in North America have historically included both wolves and coyotes as important figures in many of their traditional stories. Stories identified as Coyote stories constitute a subgenre of myths and stories in peoples from the Plains and the Intermountain West. The following excerpt from the story "Coyote Fights Some Monsters," from Mourning Dove (the pen name of Christine Quintasket, an Okanagan writer) and her "editor" Heister Dean Guie, involves several apparent canids whose identities are not obvious:

> [Coyote's] way led past a cave . . . the home of Kika-waupa—
> Dog—who was a big and ferocious monster. Dog rushed out
> of the cave. Coyote ran. Coyote stumbled, falling into a mole
> hole, and that made him think of his faithful wife, Mole. Making
> himself small, he crawled into the hole, and there was Mole. . . .
> "Make your underground trails," Coyote said. "Make many of

your tunnels. Hurry!" Mole . . . dug fast, for Dog was digging to get at her husband. She dug many tunnels, as Coyote had ordered. Dog soon uncovered Coyote, who then resumed his usual form, and said: "Wait, Kika-waupa! Do not kill me yet. Let me smoke my pipe first." Dog did not object, and Coyote smoked. As he sucked on his pipe, Coyote spoke to his squas-tenk'. It gave him an armful of stones. Hitting Dog with a stone, he ran. Dog howled with pain and rage, and set out after him. Dog stumbled over one of Mole's mounded tunnels and fell, and Coyote hit him with another stone. Dog did not know that Mole had been busy changing the ground there, and every time he came to a tunnel he stumbled and fell, and each time he went down Coyote hit him with a stone. . . . In a little while Dog was so tired and bruised that he could not take another step. Then Coyote finished him, and out of the monster's body ran a small dog, its tail between its legs. "You shall be the most faithful animal the New People will have," said Coyote to the little dog. "Even old men and old women will own you. You will both fear and like your owners. Never must you attack a stranger unless the stranger treats you badly." Coyote left the little dog. (1990, 43)

This story provokes numerous questions. First, the ferocious monster described as "Dog" could represent a wolf or another form of early wolf-like canid (perhaps even the dire wolf, *Canis dirus,* which was contemporaneous with the first Americans). As we established in chapter 7, many American tribes regarded wolves as their companions and used the same word to mean both dogs and wolves. Translation is a particularly thorny issue in the case of Mourning Dove, who is known to be one of the most heavily edited of Native American authors (Brown 1993). Her works were changed considerably by her editor, L. V. McWhorter, which is likely the case in this story because another monster she identifies in this tale is Horse, which marks the narrative as clearly postcolonial in nature. Scholars argue:

McWhorter enlisted a close friend, Yakima newspaperman Heister Dean Guie (1896–1978), to help with the shaping of Mourning Dove's traditional stories. Guie's wife, Geraldine

(1897–1994), happened to be one of the first graduates of the University of Washington's anthropology program, and her methods may have influenced many of the book's editorial decisions. Because Guie envisioned the collection as a series of children's bedtime stories, all mentions of sex and violence were eliminated, and most of the legends were simplified and shortened. Mourning Dove and McWhorter, who remained active in the editing process, removed moral points, "superstitions," and creation stories that might bring ridicule from a white audience. When *Coyote Stories* was published in 1933, it included editing credits to Guie and McWhorter. . . . Many of the stories as published were unrecognizable to the Colville-Okanagan elders who originally told them. (Nisbet and Nisbet 2010)

This suggests that the word *dog* might originally have had some other meaning but was translated as such because of European influences and Euro-bias favoring dogs rather than wolves.

A second question is whether the "small dog" that runs from the monster's body is a metaphor to describe the way Indigenous groups understood wolves to be the ancestral form of later, smaller "domestic" dogs. This seems likely because this tale contains a repeated theme of "monsters" being changed into more malleable, presumably domestic forms. Few Indigenous stories describe wolves as ferocious beasts, however, so as indicated above, this story might have changed after editing by Europeans, who brought with them a fear of wolves (McIntyre 1995; Schwartz 1997; Coleman 2004; Pierotti 2011a).

There were clearly differences at contact between Euro-Americans and Native Americans concerning their understanding of the evolutionary relationships between species, with Euro-Americans emphasizing separate creation events for all life-forms and native peoples stressing relatedness (see chapter 5 in Pierotti 2011a). These same scientific and philosophical concepts are involved in contemporary American debate over the empirical basis of evolutionary biology. *Creation* and *relatedness* lie at the heart of Indigenous understanding of the nonhuman world and its relationship to *Homo sapiens*.

The Western monotheistic religious tradition posits a creator that is human in both form and thought. Because this creator is assumed to be human, or at least humanlike, it is supposed to have human mental and psychological limitations and human values. Many followers of the Western philosophical tradition assume that if God does not think like them, then he cannot think at all, and therefore does not exist (Pierotti 2011a). This conundrum reveals the conceptual limits of the Western philosophical tradition, and also why fundamentalist Christians resort to intelligent design, which seems to assume that the creator functions as a master engineer (Petto and Godfrey 2007; Pierotti 2011a).

In contrast, Native Americans assume a creator that is not human or even remotely humanlike in its thought patterns. This entity would not see humans as superior to other life-forms because all life-forms are its descendants. In Indigenous cultures it seems likely that the concept of "creation" or "origin" refers to a series of events that are not located in specific periods of time, but instead are linked to a particular physical location where environmental conditions changed, forcing a culture to redefine itself and develop new traditions and ways of coping (Pierotti 2011a).

We are discussing cultural traditions that were being established at the end of the Ice Ages, when conditions were highly variable, and humans had to contend with massive flooding, unpredictable freezes, and intermittent warming, in addition to having to eke a living by hunting and gathering while animal populations changed dramatically. Deglaciation in the Northern Hemisphere began about 20,000 YBP (Clark et al. 2009), which corresponds to most recent evidence concerning the timing of the peopling of the Americas (Dillehay et al. 2015; Gibbons 2015). Climate change has been a major driver of changes in population size among both humans and nonhumans over the last 50,000 years (Pennisi 2004; Shapiro et al. 2004; Lorenzen et al. 2011; Gibbons 2013).

If creation stories deal with responses to changing environmental conditions, this suggests that Indigenous cultures were attuned to variability in the environment, and that such cultures were capable of making major adjustments in lifeways in order to accommodate environmental changes. "Tribal systems . . . are not static in the sense that they do not allow or accept change. Even a cursory examination of tribal systems will show that all have undergone massive changes while retaining those characteristics of outlook and experience that are the bedrock of tribal life" (P. G. Allen 1986, 63).

An analogous version of this interpretation can be found in the Jungian tradition, where it is assumed that "creation myths . . . represent unconscious and pre-conscious processes which describe not the origin of the cosmos, but the origin of man's awareness of the world" (von Franz 1995, 5). Thus, the concept of "creation" can be applied to the origin of particular forms of human awareness or to the beginning of a new cultural tradition, which are fundamentally the same thing (Pierotti 2011a). This is the basis of what we call myth—not fabrications but memories embedded in stories concerning events that happened long before our time, events that Lakota scholar Joseph Marshall III describes as "beyond memory."

The origin of new cultural traditions in response to changes in environmental conditions is analogous to the formation of new species, a phenomenon also typically assumed to result from drastic new changes in environmental conditions (Pierotti 2011a). This has implications for the theme of this book, because both human social cultures and wolf social behavior and evolution changed as a result of their partnership, which took place during this time of major global environmental change. One reason that humans may have been driven to partner with wolves when they spread northward was because, being recent émigrés from tropical Africa about 40,000 years ago (Harris 2015; Shipman 2015), modern humans were inexperienced in dealing with cold and changeable climates. Wolves, however, were very experienced in such environments, and may have been much more sociable to and accepting of the new arrivals than the other hominids in Europe and Asia, such as Neanderthals or Denisovans.

With regard to North America, nonhumans attributed with creator status in Indigenous American cultures are frequently the most intelligent and social creatures of their particular ecosystem or community, such as wolves, coyotes, ravens, and bears (Pierotti 2011a, 2011b). These creatures are the most powerful and knowledgeable entities within particular local ecosystems, and their existence provided both a livelihood for humans and a framework within which cultural traditions could develop. The Western tradition of viewing human history as independent of place and ecology is not part of the conceptual framework of Indigenous peoples of North America (Deloria 1992; Basso 1996; Pierotti 2011a, 2011b). For Indigenous Americans, religion, or at least spiritual traditions, is used to code ecological knowledge, which connects Indigenous

peoples to the place from which they came and establishes their relationship with other species.

The wolf was a species of great cultural and spiritual significance to
many Indigenous American peoples (Schlesier 1987; Marshall 1995;
Barsh 1997, n.d.). Wolves existed throughout North America, lived in
family groups, and were not strong or swift enough to kill large prey
alone. Wolves served as models for the concepts of community existing at
both the single-species and ecosystem level (Bruchac 2003, 159). Like
humans, wolves proved capable of associating with and maintaining
cordial relations with other species, such as ravens (Barsh 2000). Western
scientists studying wolves recognize that native people have far greater
knowledge of the behavior and ecology of wolves than does Western
science, and they turn to native people to help them in their study of these
animals (Stephenson 1982).

As discussed in the previous chapter, numerous tribes view wolves as
important teachers who helped humans in their efforts at hunting (Fogg
2012; Fogg, Howe, and Pierotti 2015). Some tribes view Wolf as the creator of
all things, including the people themselves. "For some tribes, the First
People include a noble, heroic figure, such as the Wolf among the Comanche,
who foresees the coming of humanity and plans a perfect, ideal world for
them—until his brother coyote enters the scene as marplot" (Bright 1993,
20). In these cultures, Coyote is often described as the trickster figure. After
the introduction of dogs by Euro-Americans, Dog was recognized to be
Wolf's cousin by the Comanche, and since coyotes and wolves are taboo as
food, the same must be true for dogs (Wallace and Hoebel 1948).

The manner in which tribes such as the Shoshone and Comanche
see the wolf and the coyote reflect their understanding of balance in
nature. Wolf (Pia Is'a, pronounced *Pee'a Eesha* in Comanche and *Bee'a
Eesha* in Shoshone) is seen as the creator figure, and Coyote (Is'a) is his/
her little brother who, like many younger siblings, is constantly trying to
improve upon its older sibling's efforts (Ramsey 1977; Buller 1983;
Vander 1997). The role each plays contrasts with that of the other;
however, it is important to note that Coyote is not regarded as an evil
figure. He/she is thought of more as a mischievous figure and as a sort of
subcreator, which is why he is described as the little brother of Wolf and
not as an opponent, except possibly in the intellectual sense. "Wolf was
wholly beneficent; his acts of original creation made all things perfect and

good. Coyote, the mischievous Till Eulenspiegel of Shoshonean folklore, was the spoiler of all things, however. His was the role of the transformer who undid the good works of his big brother. He brought hardship, travail and effort into the lives of men. He represented the force of Evil as we [Euro-Americans] see it—and yet the Shoshones in no way thought of him and his relationship with Wolf as a conflict of good and evil. Coyote was not bad, he was no more than wantonly mischievous" (Wallace and Hoebel 1948, 193–94). Thus, in the tradition of the Nuhmuh/Newe (Shoshone and Comanche peoples), Wolf was considered the creator figure who created a perfect world (A. M. Smith and Hayes 1993; Harney 1995, 26; Papanikolas 1995), although at times Wolf loses patience with his creation, as shown in the following story, "Tracks of the Creator":

> It is said that the [Paiute] Creator, Gray Wolf (*Numuna*), burnt everything in the old world. . . . Gray Wolf talked with the Sun, "There should be a flood" . . . and the Mountains were covered with Water. Then Gray Wolf went with his woman far across the water.
>
> After a time, it is said, the water began to dry up, the mountains appeared, there were banks and shores again. Then the Sun said to Gray Wolf, "You should make children." "Yes," said Gray Wolf, and he created pine trees, juniper trees, aspen trees, cottonwood trees, willows, springs, deer, otter, beaver, trout, buffalo, horses, mountain sheep, bears.
>
> When Creation was finished, Gray Wolf's children began to do wrong; they fought amongst themselves. . . . Gray Wolf became angry, and kicked them all out. He decided to go south; he said, "My children are not going to see me again!" Then his wife cried, "But my children are here!" But they went down to the water anyway, it is said, and walked away over its surface.
>
> Gray Wolf and his wife came to a tall mountain, with a pine-covered summit. He said, "I am going in there; afterwards my children will see my tracks going in. Here I have come and left my tracks; Nuhmuhnuh will see them and so will white men." So it was. (Ramsey 1977, 231)

One point of interest in this story is that the word *Numuna* (also spelled *Nuhmuhnuh*) can be translated as either gray wolf or as human being (Ramsey 1977, 231), which suggests that humans and wolves were considered to be equivalent as persons, which follows from their ecological equivalency within the Numic (Shoshonean) tradition. Another interesting theme is the echoes of death and reincarnation, which makes sense because according to the Numic peoples, Wolf and Coyote frequently argue about how the world should function, with Wolf desiring an idealized world in which death is only temporary, childbirth is easy and pleasant for women, and winter does not exist. In contrast, Coyote thinks death should be permanent, childbirth should be difficult, and hardships and cold weather should be regular aspects of human experience (Lily Pete, in A. M. Smith and Hayes 1993, 3–4; see also Ramsey 1977; Harney 1995; Papanikolas 1995). An example of this conflict is seen in this story told by the late Shoshone elder Corbin Harney:

> You see, the coyote and the wolf were talking long ago. Wolf was arguing that we should all look alike, the rocks should be the same, the sagebrush the same, the humans the same, and all the living things on this planet should be the same. We should think alike and act alike and so forth.
>
> But Coyote always said, "No, we should be all be different. We shouldn't look alike at all." And so today we look around us and nothing looks alike. Rocks are not alike. Humans are not alike. This is the root of why we don't believe in each other. It's just as Coyote said. "There's no use believing in just one thing. Let's not believe it. Let's all disagree, and everybody believe in different things."
>
> That's why I always say, it's easy to believe the bad things first, but the good thing is harder to believe and harder to come by. As Wolf said, "It's going to be really hard that way, because what you're saying is, let's not believe in each other." So today, what Coyote said is what we've got. (1995, 26)

This story shows that, although Wolf is perceived as the creator, Coyote often wins arguments, which explains why he/she is regarded as a co-creator in many ways. Wolf is the Pia Apo, Big or Great Father (in Shoshone pronounced Bee'a Apo, with the o voiced as a soundless expression of breath; Miller 1972), whereas Coyote is Tei Apo, Little Father (Vander 1997, 81). The story also sheds light on why there are more

Coyote than Wolf stories told—as a creator of conflict, Coyote is the more interesting character.

Bivo, a Wind River Shoshone, stated that during the Sun Dance Ceremony, "prayers were addressed to the Sun and Apo [the Father], the latter being Coyote's elder brother Wolf" (Lowie 1909, 199). Nonetheless, Coyote acts as the corrective; although children are taught to emulate Wolf and view Wolf as a much more sympathetic figure than Coyote, it is Coyote who presents the more realistic view of how the world truly functions. Coyote is the realist and Wolf the idealist—and in the long run, it is obvious who will usually carry the day.

Although Coyote generally wins these arguments, the Shoshone regard Coyote as highly conflicted over the real consequences of his/her beliefs, as seen in "Controversy over Death":

> A long time ago people never used to die. Coyote's brother, Wolf said, "When people die they will get up again after two days." Coyote didn't like that. Coyote said, "When we die, we should die forever." Wolf didn't like that. Coyote kept on asking his brother why the dead should get up. . . . After that Wolf wished that Coyote's son [Magpie] would die. . . . After Coyote's son died, Coyote went to his brother's place and said to him, "Raise my son to life after two days." Wolf didn't answer for a long time. Then he said, "You, Coyote, said that people should die forever." Wolf told him to burn all his clothing, and cut off his hair and burn it. Wolf told him that dying forever is what Coyote wanted in the first place. If it weren't for Coyote there would be too many people now.
>
> After Coyote burned his clothing he lay down flat on the ground and looked up at the sky. Soon he saw lots of [crows] up there. . . . They belonged to his big brother, Wolf. Coyote wished that one of them would fall down. . . . He saw one coming down. . . . Coyote caught it, [and] tore the crow to pieces . . . because he was angry at Wolf. Coyote had a funeral service for his son. He sang all night. (Commodore, translated from Newe by Lily Pete in A. M. Smith and Hayes 1993, 3)

Although it seems simple on the surface, this story is both profound and rooted in reality. It features an existential debate about the nature of life and death, and the one who wins the argument quickly comes to regret his "victory." Coyote's response is very human—who among us has not said, "Oh, no, it can't be true" when we lose a loved one? Both Wolf and Coyote are seen as fallible and vindictive, even if Wolf is portrayed as an idealist. From an ecological perspective, the choice of the targets of their respective vengeances are interesting, because magpies associate with coyotes in the same way that crows and ravens associate with wolves (see chapter 2). Despite the sadness in this tale, the Newe recognize that "if it weren't for Coyote there would be too many people now," which reveals the truth behind the troubling reality and shows that the people are more aware of the risks of human overpopulation on available resources than are most contemporary Europeans, Asians, Africans, and Euro-Americans. Numic peoples were known to practice family planning hundreds of years before this became a concept in other cultural traditions (Wallace and Hoebel 1948).

After Wolf becomes discouraged with the world, as described in "Tracks of the Creator," the subsequent role assigned to him is to escort the dead to the next world (Vander 1997). Halfway up in the sky to Wolf's house, the *mu'gua* (soul) turns into a *dzo-ap,* or ghost. The former soul takes on a new form, becoming a whirlwind. A point of no return has been reached: death is final (Lowie 1909, 226).

There is debate among the Shoshone concerning whether Wolf as Apo (Father) means that he is the equivalent of the Christian God the Father, or if he is simply Wolf, the equal of Sun and the father of the Newe traditions (Vander 1997). Indeed, the addition of Christianity to Indigenous cultural traditions seems to have thrown many of these spiritual concepts into confusion. In older Shoshone tradition, Wolf is the wise older brother, Coyote the mischievous younger brother, and both serve as creator figures. Some Wind River Shoshones believed that there was a "Father" who created both Wolf and Coyote, but he made them as his "brothers" (Vander 1997, 81)—a confusing idea, to say the least. More traditional Shoshone simply argue that Wolf is the "Father" (Bia Apo). This "father" terminology allowed Christian missionaries to sow confusion, and in consequence, many people conflate their own spiritual traditions with those of Christianity. In this context, it is interesting that the Teton Lakota address Wolf as "grandfather" (Marshall 1995), implying that Wolf is part of their direct spiritual ancestry while avoiding the "Our Father" ambiguity by establishing a less filial relationship with Wolf.

To Numic peoples, however, Wolf is obviously linked to "the Great Spirit" as the elder, more powerful entity. Wolf is Pia (Bia) Apo, the "Great Father," and "Coyote is God's younger brother" (Vander 1997, 81). To quote a Wind River Shoshone, "We only knew the name Our Father (*Bia apo*) to whom we prayed, but since then we have learned his name from the white people, God" (Shimkin 1939, 26).

The figure of Bia Apo is a complex entity to the Shoshone. Some "think about him as the whole sky, he covers the whole world" (Vander 1997, 80). Others associate Bia Apo with the sun, but they are emphatic in stating that they do not worship the sun—they pray "through" it. As stated above, Bia Apo is Wolf. All this is made more complex by the additional concept that Bia Apo's realm is the western sky, the direction of the setting sun, which is also the final destination of the dead (80). This returns us to the story detailed above: after creating the world, Wolf became disheartened by the failure of his creation to live up to his expectations, and retired to serve as the escort of the dead to the next world.

Among contemporary Shoshone there is a "confusing array of primal kin relations embodied as animal characters in myth. This includes father, older brother, younger brother, Sun, Wolf, and Coyote" (Vander 1997, 80). Regardless of this complex cosmology, however, the conception of Wolf being the "older brother" to Coyote (Vander 1997) is almost certainly based on real-world observations of the ecological and social relationships that existed between these two species as co-predators on the Great Plains and in the Intermountain West (Pierotti 2011a).

As we have discussed in other chapters, humans and wolves share a long ecological relationship and a coevolutionary history (Morey 1994, 2010; Schleidt 1998; Sablin and Khlopachev 2002; Schleidt and Shalter 2003; Ovodov et al. 2011). In this context, it is crucial that wolves, and not coyotes or jackals, are the direct ancestors of dogs, which are the closest companion animals of contemporary humans (J. Clutton-Brock 1984, 1995; Morey 1994, 2010; Vilà et al. 1997; Derr 2011). The reason that Wolf is so important to many North American Indigenous peoples is that humans and wolves, both wild and domestic, shared cooperative foraging relationships well before either species emigrated from Asia into North America (Morey 1994; Vilà et al. 1997; Schleidt 1998; Germonpré et al.

2009; Ovodov et al. 2011). What is interesting is that when humans arrived in North America, they also found indigenous wolves and quickly established, or reestablished, similar sorts of cooperative relationships.

This line of thought is further illustrated by the observation that in Indigenous belief systems, creators are typically nonhuman entities embodied as a species of animal recognized as important to the local ecological community (Pierotti and Wildcat 1997; Pierotti 2011a, 2011b). What is important about this distinction is that the Western monotheistic religions all work to separate humans from the rest of nature (Gray 2002), whereas Indigenous cosmological thought insists that we recognize that humans are part of nature, irrevocably tied to the other species that have come down through the millennia as prey, rivals, companions, or some combination of these three roles (Pierotti 2011a, 2011b). This reinforces basic Indigenous concepts of relatedness and connectedness, because significant nonhuman species are considered to be the originators of cultural traditions and perhaps of modern human beings themselves. This is exemplified by the treatment of Raven as the creator figure in cultural traditions of the Pacific Northwest and Alaska (Nelson 1983; E. N. Anderson 1996), by the treatment of Wolf and Coyote as co-creator figures in the Plains and the Intermountain West of North America (Buller 1983; Marshall 1995; Vander 1997), and by the Australian Aboriginal recognition of Dingo as the entity that makes them human (Rose 2000, 2011).

The ecological—and especially the socially dynamic—similarities between humans and wolves can be interpreted from a cultural evolutionary perspective as evidence of close social and ecological relatedness (Pierotti 2011a, 2011b; Fogg, Howe, and Pierotti 2015). Part of the Western scientific tradition argues that ecological similarity can lead to organisms being considered as members of the same taxonomic unit under the *Ecological Species Concept,* which defines a species as a lineage or a closely related set of lineages that occupy an adaptive zone minimally different from that of any other lineage in its range and which evolves separately from all lineages outside its range (Van Valen 1976). This can be recast as the set of organisms that occupy a single ecological niche (Pierotti 2011a). The ecological niche is comprised of the set of resources and habitats exploited by a species, or a group of species if their niches overlap to a considerable degree. Following this logic, the tendency of wolves and humans to work closely, both within and between species, in foraging activities and prey choice and behave in a similar manner, led to their being considered to be one another's closest relative by many Indigenous peoples of the Great Plains (Marshall 1995; Fogg, Howe, and Pierotti 2015).

Wolves were the most visible social and intelligent predators in the environment of the Great Plains. In the Pacific Northwest, the most intelligent and social animal that is highly visible and interacts with humans on a regular basis is Raven (Pierotti 2011a, 2011b). Ravens differ ecologically from wolves, being generalized omnivores and predators on small animals as well as scavengers on carcasses of salmon and large mammals; however, like wolves, ravens are also monogamous and live in extended family groups (Heinrich 1999; Pierotti 2011a). Raven is regarded as both a creator and trickster figure by tribes in the Pacific Northwest (Nelson 1983). As we pointed out in chapter 2, however, wolves and ravens are also close ecologically and socially, considered to be friends to the degree that ravens are referred to as "Wolf-Birds" (Heinrich 1999).

The nature of hunting relationships and the prey exploited are very different on the American Plains and Intermountain West than in the dense forests and colder climate of the Pacific Northwest and the Arctic. Ravens are not considered to be close relatives and equivalents of humans in the same way that wolves are. Wolves are considered to be creators in areas where humans and wolves exploit similar types of prey, such as bison, pronghorn, elk, mule deer, and bighorn sheep. Ravens are considered to be creators in areas where they help humans to locate prey such as caribou (Brody 2000; C. L. Martin 1999).

Thus, it appears that nonhumans regarded as creators by the Indigenous peoples of North America are typically those species that are most intimately linked to the ecology and lifestyles of the human cultures in a particular ecosystem. In the northern forests where bears helped humans learn about plant foods, they were sometimes considered creator figures (Pierotti 2011a).

To conclude, it is likely that in many ways referring to Wolf as a creator is not all that different than considering wolves to be important teachers, especially of hunting and social skills. As both creator and teacher, Wolf has been part of the peoples' cultural traditions from "the other side of memory." When the first humans crossed Beringia to enter the land now called North America, they were almost certainly guided by ravens and accompanied by wolves. These two species help code the memories of various peoples and cultural traditions as they adapted to the new environments they encountered. Even more important, these two species helped humans survive in lands about which the nonhumans knew far more than did the humans. Since most of these survival skills involved hunting and scavenging, it is hardly surprising that the most intelligent and social hunter and the most intelligent and social scavenger are the two nonhuman species most often recognized as creators by the peoples of North America, or that they also seem to be each other's closest allies. Wolves, ravens, and humans shared kills or fed from remains of kills made by the others (Schlesier 1987; Marshall 1995; Hampton 1997). All live in close family groups dominated by a mated pair of adults. Following our arguments and those of Schleidt and Shalter (2003), wolves may have been a major force that shaped human evolution, creating a territorial, cooperative, group-living species with strong family bonds from an intelligent ape which had been none of these (Rose 2000, 2011; Schleidt and Shalter 2003). One major reason that Wolf (and Dingo) are viewed as creator figures is that they may literally have made us the humans we are today.

WHAT AND WHY IS A TRICKSTER?

In no culture on earth is Wolf ever characterized as a trickster, although in some contemporary Siberian stories Wolf may be a buffoon and the object of scorn. We have referred to both Coyote and Raven as "tricksters" without discussing what that means to the cultures who employ such concepts. Our goal here is to attempt to define what a trickster is in an ecological and social context, and further, why this term may not be appropriate when discussing entities based on actual living creatures.

The standard definition of tricksters tends to follow that developed by Paul Radin: "Trickster is at one and the same time creator and destroyer, giver and negator, he who dupes others and who is always duped himself. . . . He knows neither good nor evil yet he is responsible for both. He possesses no values, moral or social . . . yet through his actions, all values come into being" (1972, xxiii). One interesting facet of this definition pointed out by L. Hyde (1998) is that this description definitely does not fit the Christian devil or Satan, because the trickster is duped as often as he dupes, and is not the personification of evil but instead "knows neither good nor evil," and his actions lead to the values accepted by society.

Despite clichéd attempts to identify tricksters with Satan, tricksters have little to do with evil, even if their purported actions often seem to be harmful from a European perspective. Some scholars (e.g., Ricketts 1966) have argued that all tricksters are to some degree "culture heroes," and that "in the most archaic hunting cultures of North America trickster and hero roles are always combined in the same figure" (329). More interesting is that Ricketts further argues that the trickster is "man himself, while his actions as related to the myths disclose man transcending himself" (344). It is clear from the Shoshone tradition that despite his personal conflicts, Coyote recognizes the difficulties of life in a complex world and comes up with solutions, even if these are initially unpopular.

For example, Coyote's insistence on the permanence of death flies in the face of Wolf's utopian naïveté, but it is obvious that Coyote's interpretation reflects reality. Coyote is not a mystical figure—instead, he is a sometimes gross, frequently selfish, often lazy, but ultimately rational being: in other words, Shoshones as they recognize themselves to be but are either too polite or too embarrassed to acknowledge openly. They hide in the disguise of Coyote, an intelligent small predator clearly like Wolf, but not Wolf, that is often too smart for its own good.

The image of the idealistic but somewhat petulant Wolf is contrasted to the iconoclastic but practical Coyote. Wolf wants things to be perfect and gets frustrated when they turn out not to be as simple as she/he imagined. This is why in the Numic philosophical tradition Wolf is given charge over "heaven," whereas Coyote must deal with the reality of Earth. Thus, Coyote cannot be an idealist; the physical world is too real and cannot be ignored. One interesting debate that emerged from the Ghost Dance, a spiritual movement brought about by the Paiute Wovoka as a response to the destructive power of the invading Euro-Americans dealt with practical matters. As part of his preaching Wovoka contended that if all Indians participated in the Ghost Dance, this would bring back the buffalo as well as all of the people killed by whites. This point required clarification for some eastern Shoshone and Lakota, who wanted to know how all these people would be accommodated, and who would feed them? This reveals that these people remain practical children of Coyote, whereas Wovoka, invoking his inner Wolf, contended that the creator would increase the size of the earth, and therefore all these people could live without increasing population density (Vander 1997).

Once Europeans arrived in North America as residents, a number of the tribes began to associate them with the concept of trickster, apparently because their motivations seemed unclear and because they tended to consider as important things that Indigenous people thought were marginal or peripheral (L. Hyde 1998; Ballinger 2004). They were obviously human but seemed trapped between adulthood and childhood, quite immature in their attitudes and relationship with truthful speaking. This concept seems to have been intended as a mild reprimand because trickster stories are often told to show how it is proper to live (or not live), as with the discussion about death and reincarnation. Nonetheless, dealing with Europeans was deadly serious, even if it had humorous overtones.

In contrast to Radin's definition of tricksters (1972), a more subtle and indigenous definition is provided by Ramsey: "The Trickster is an imaginary hyperbolic figure of the human . . . whose episodic career is based on hostility to domesticity, maturity, good citizenship, modesty, and fidelity . . . given to physical disguises and shape-changing; and who in his clever self seeking may accomplish important mythic transformations of reality, both in terms of creating possibility and of setting human limits. From a structural standpoint, tricksters are important mediative figures" (1999, 27–28). It is important to emphasize that Ramsey's use of the term *mediative* implies a dynamic interposing of the mind between polar opposites, allowing it to hold on to both, as in Yeats's claim that he could "hold in a single thought reality and justice." This does not mean "compromise or reconciliation"; rather, it is a continuing process of the mind, not a transitional step toward a conclusion (29).

Lévi-Strauss (1967) argued that nonhuman tricksters among western tribes tend to be scavengers and omnivores, such as coyotes and ravens, who occupy an ecological mediating position between herbivores and carnivores and are thus "in between" in terms of subsistence strategies. This seems clever, especially because humans are also omnivores, and at least in their early history probably functioned as scavengers; however, it is unclear what it means to mediate between carnivores and herbivores. Omnivorous animals eat some fruit and may scavenge in refuse heaps, but in their natural systems they are basically carnivores, and not in any sense herbivores.

Ramsey (1999) mentions Lévi-Strauss's analysis; however, he makes the more compelling argument that it is probably best not to consider

Indigenous American traditions in terms of the romantic inclinations of Western writers like Yeats and Blake. Indigenous Americans seem to regard tricksters more as collective figures who are unselfconscious and unheroic but hold together the various polarities of their experience, such as grief over loss combined with recognition that death is part of life or (if you are a teenager) that sexual urges need not be expressed each time they are felt if you are to function as a member of a mutually respectful social group.

In Ramsey's view, the role of tricksters is "to provide important human identification through the adventures of an ambiguous creature who is somehow intermediate between human and nonhuman" (1999, 30). This state of being both human and nonhuman is problematical, however, because it sets the "trickster" at odds with the nonhuman natural order. Such beings abuse and in turn are scorned by other nonhumans because they all too humanly waste natural goods, and are greedy and perverse. They also are capable of a range of deeds that are "not allowable according to the canons of idealized human nature," even if such actions may be natural for nonhumans (31).

This is an area that we also find problematic. If a pair of coyotes has sexual intercourse "in public," this is of concern only to humans, irrelevant in the natural world. Similarly, if a female wolf or dog kills puppies produced by another female within her pack, she is clearly conflicted behaviorally (McLeod 1990; Marshall Thomas 1993), but this is not an immoral or illegal act. It fits within the framework expressed by the Shoshone Commodore, that if death were not a reality, there would be too many mouths to feed.

What makes the trickster more conspicuously human in its essence is that its primary role is as "mediator between the antinomies of tribal good citizenship and individual self-fulfillment" (Ramsey 1999, 32). Trickster greed or perversity is used to inform tribal or band members of what types of behavior are unacceptable. This is clearly important in maintaining social norms without having to resort to having a "police force." Such actions, however, have nothing to do with actual living animals, such as coyotes or ravens.

One of the key functions of trickster stories is their role in the training of the next generation. "Another of the trickster's interlocking mediative functions is to 'get between' children and their unverbalized fears and preoccupations about themselves, especially about their growing bodies in relation to their developing selves" (Ramsey 1999, 37). These stories reduce guilt and anxiety at the individual level while masking important lessons about social functioning within a humorous context that is shared by the entire community. Such stories can function as correctives by reference to the story, as when a young person violates a social rule or otherwise gets into trouble (Basso 1996). The trickster provides a "type specimen" of an undeveloped mind, offering historical examples of how uncouth and infantile it was around here in the beginning time, which simultaneously points forward toward the sophisticated present occupied by the "People" in their current incarnation. For confirmation of how it was in mythic times, all the children need to do is look closely at themselves as they try to avoid such errors (Ramsey 1999, 38).

It is interesting that Indigenous Americans assign the trickster role to nonhumans. In contrast, tricksters in European traditions tend to be gods, such as Hermes or Loki, or other mythic figures—for example, Prometheus, a Titan from the generation that preceded the gods of Olympus. This means that Indigenous American tricksters need not act out of noble altruistic "promethean" motives; they can be responsible for major changes by accident—as in the case of Coyote's arguments about death—or through selfish motives—such as Coyote demonstrating the dangers of incest or Raven releasing the sun because he is having difficulty hunting in the dark.

To accept that your world is largely the work of a trickster is "to know and accept it on its own terms" (Ramsey 1999, 41). It's not perfect, but given the being who created it, how could it be? It's good enough, and it's ours. This shows an acceptance of nature and a corresponding belief that the way the world works does not need to be massively changed or "engineered" to make it a better place to live (Gray 2002; Pierotti 2011a). Wolf may be the entity that got things started and tried to make a perfect world, but Coyote and Raven, Wolf's younger brother and his constant companion, shaped this world into the somewhat problematic place it has become. It is instructive that in more recent times, Coyote (along with wendigos) has become a stand-in for Europeans, whose immature, greedy, and selfish behavior is seen as primitive and arrogant. The powerful white men came in to take over the role of trying to remake the world through greed-driven errors (Gray 2002).

From our coevolutionary and ecological perspective this worldview makes sense. Wolf could function as a teacher and creator to Indigenous peoples, who were willing to respect wolves and pay attention to the examples they set. Wolf is removed, however, from interacting with European culture, which is completely hostile to wolves and treats them as bitter enemies. They were never teachers or creators to these folks. In contrast, it is ecological and evolutionary fact that ravens and coyotes are two species that have thrived since Europeans invaded (Ballinger 2004). Their ranges have increased and so have their numbers as they take their generalist ecologies and learn to live in the spaces of the environments created by their fellow trickster the white man.

A COMMENT ON SHAPE-SHIFTING AND IDENTIFICATION WITH NONHUMANS

In many traditional Indigenous American stories characters are described as being able to shape-shift into a wide variety of organisms, including seals and whales (Martin 1999). Much of this material is related to confusion over shamanism and what it represents in terms of objective phenomena. One thing often omitted in this discussion is that the performances of these dances, including the regalia worn in many of them, were meant to create an atmosphere in which the dancers could identify with and even "become" the animals they would subsequently hunt. Roger Welsch describes a Buffalo Dance: "It is impossible once the dance begins to know who the heavily costumed and masked dancers are, nor does anybody guess. The [People] understand that the dancers are now buffalo . . . no longer their relatives, friends and neighbors once they put on their Buffalo Dance regalia and begin to dance. In the matrix of the costumes, prayers, music, and ritual those who were human beings earlier in the evening . . . are now buffalo" (1992, 60). Later Welsch states that "he [the Nebraska state archaeologist] tried to look into the dancer's eyes, but it was as if there was no dancer, only the mask" (74). Finally, Welsch notes, "A Buffalo Dancer is a spirit, not a human being" (67). This is why you see Buffalo, Deer, and Eagle dances, but not Wolf, Coyote, Bear, or Cougar dances, because carnivores are not hunted for food, and they are regarded as equals by humans. In a similar fashion, dancers of the Pacific Northwest, with their complex masks and regalia, can invoke the animals whose appearance they are taking on so accurately that it can be disconcerting to onlookers of European ancestry, as Pierotti knows from personal observation.

One subcategory of this topic that gets a lot of attention, at least in the media, is stories of shape-shifting into wolves. The subject has taken on a new life in the popular series of *Twilight* books by Stephenie Meyer, a Mormon housewife. The books and the movies based on them feature (along with vampires) a pack of Native American werewolves on a loosely disguised tribal reservation on Washington's Olympic Peninsula. Such tales typically derive from stories of Navajo "witchcraft" or "skinwalkers" (a regular subtheme of Tony Hillerman's best-selling novels), in which such entities are described as wearing the skin of a wolf or coyote while participating in various unpleasant activities (Kluckhorn 1944).

> The Navajo word for this practice is *adishgash*, and for the one who practices it, *adilgashii*. The Anglo missionaries and anthropologists who followed the cavalry translated *adishgash* as "witchcraft." More often one hears the word "skinwalker," a rough translation of the Navajo phrase *yee naaldlooshii*. While practically anyone with a grudge can work *adishgash*, a skinwalker is a studied and practiced malefactor who is able to shapeshift into a wolf, owl, or other noxious creature—hence the sometimes-heard name "Navajo wolf." Navajos take these concepts very seriously. In 1860 the U.S. Cavalry rounded them up . . . , and marched them 300 miles east to Bosque Redondo. Recalled as *Hweeldii* (literally, "hardship"), this event was their Holocaust. After their return in 1864 they conducted their own witch purge.
>
> "The motives for the witchcraft were the time-honored ones for all evil deeds—envy and jealousy." . . . The witchcraft was directed by a group of impoverished Navajos living near Chinle, Arizona, against their well-to-do neighbors. The victims became sick; their cattle died and their crops failed. The resulting purge claimed 40 lives before the bloodshed was put down by the U.S. Army. It seems that being too successful materially can disturb the collective sense of *hozho* [balance and harmony]. At the most fundamental level, a "witch" in Navajo tradition is a person who acts out of selfish motives.

The cultural imperative to share resources is so strong that a relative who asks to borrow your new truck cannot be turned down, even if you are pretty sure he's going to buy booze, drive drunk, and wreck it. Better to let him crash than disturb your family's *hozho*. (Brenner 1998)

A more reasonable alternative form of "shape-shifting," which might derive from experiences like those described in the previous chapter, is acquiring wolflike tendencies as a result of prolonged exposure to a wolf pack, typically after having lived among them. In such stories wolves are idealized, and becoming one is considered to be a noble transformation. Within these stories the character that will eventually shape-shift is wronged or abandoned by other humans. In the story "The Forsaken Boy; or, Wolf Brother" from the Mohawk and Onondaga Nations, older siblings leave their younger brother in the woods to starve. However, the boy survives on leftover scraps intentionally provided by wolves, eventually learning to live among the wolves. When the older brother returns to the family's land and finds the younger brother alive, the boy's transformation begins. The harder the older brother tries to reach his younger brother, the faster the transformation occurs, until finally the younger brother exclaims, "I'm a wolf!" and runs off.

This category of story represents a variation on the wolf as helper stories discussed in chapter 7, but with a twist that suggests another way in which wolves might be regarded by the tribe. In this context, the wolf is viewed as a long-lost brother, which generates a reciprocal relationship between the two species. In the story, the wolves leave some of their kill for the small boy so that he may survive, and his family reciprocates by feeding the wolves, not knowing which one may be their missing sibling. This reflects a tradition in many tribes of leaving parts of a successful hunt for the wolves and other species out of respect.

One consequence of the stories involving shape-shifting in Indigenous traditions is the misleading comparison of these stories to European myths about werewolves (Colshorn and Colshorn 1854; Summers 1966), a confusion exploited by Stephenie Meyer and the film-makers of the *Twilight* series. The motivation for stories about humans transforming into wolves was very different for Europeans and Native Americans. For Native American tribes, with the exception of Navajo skinwalkers, who are never regarded as actual wolves (Kluckhorn 1944), transformation stories were used as a way to show similarities between the lives of humans and wolves. In European traditions the stories were used to scare people by showing the consequences of forming relationships with the devil.

Like Indigenous Americans, medieval Europeans accepted animal spirits, recognizing both bears and wolves as ancestors and totems. As did Indigenous American hunters, they would put on regalia, including animal skins, "imitate [an animal's] cries and behavior, dance actively until they enter a state where they abandoned their human state, and finally reach the spirit world" (Pastoureau 2007, 45). The Roman Catholic Church, however, attacked relationships between Europeans and the nondomestic animals with which they had cultural and emotional ties. When the church was solidifying its hold on European imaginations, it demonized European shamanic traditions and rituals in which animal spirits were invoked—for example, the berserker tradition, in which young warriors became dual creatures, imitating the gait and sounds of bears; some believed that they were metamorphosed into bears in preparation for battle (45). Interestingly, many of these warriors were actually called "wolf-cloaks" because they wore wolf skins when going into a trance or into battle.

The bear terrified the church of the Middle Ages, which characterized it as the most dangerous of animals and even as a creature linked to Satan. This fear led to a war on bears and wolves that lasted for several centuries, even leading to trials in which animals were tried, and always convicted, of various crimes (Coates 1998; Pastoureau 2011), which in turn led to organized massacres of bears.

After the bears were almost all gone, the church turned to wolves with frightening efficiency. Such traditions emerged from the odd argument, endorsed by Augustine, Aquinas, and Descartes, among others, that animals lacked souls and that fostering confusion between human and animal nature was an abomination. Wolves were not feared by Europeans during the "Dark Ages"; however, by the "Enlightenment" wolves had become the ubiquitous enemy of "civilized" European society. Wolves were exterminated in England during the sixteenth century, in Scotland by 1684, in Ireland by 1770, in Denmark by 1772, in Bavaria by 1847, in Poland by 1900, in France by 1927, and over almost all of the

United States by 1950 (Alaska did not become a state until 1959). This extermination in the United States did not involve simple shooting; treatment of wolves by European immigrants was exceptionally vicious (Coleman 2004; Pierotti 2011a). We have sometimes had students ask why wolves are hated and feared by Americans. The answer is obvious: the emotions emanate from the church, which explains why the most anti-wolf segments of American society often come from strong religious backgrounds.

CHAPTER NINE

The Process of Domestication

TAME VERSUS FERAL AND DOMESTIC VERSUS WILD

Fanciers select their horses, dogs, and pigeons for breeding when they are nearly grown up: they are indifferent whether the desired qualities and structures have been acquired earlier or later in life if the full grown animal possesses them.

Charles Darwin, *The Origin of Species*

Eighteen years ago Pierotti and his family agreed to share space with a feral cat (*Felis sylvestris,* domestic variety). She was a kitten, or more accurately a juvenile (about three to four months of age), when she showed up. She readily socialized to humans, but she was not a house cat, preferring to live in our barn and hunt house sparrows (*Passer domesticus*) as well as various species of small mammals, including rabbits. Since she lived in the barn, her primary social bond was to our small group of domestic donkeys (*Equus asinus*). She joined the donkeys when they wandered our fields, often riding on their backs—a good vantage spot for hunting prairie voles (*Microtus ochrogaster*) and cotton rats (*Sigmodon hispidus*). She also rode on the donkeys when the ground was wet or snow covered. On cold winter nights, she would jump onto a donkey's back and curl up to sleep.

Two sets of behavior convinced us that this cat's primary socialization was to the donkeys rather than to humans. First, she had established a good relationship with our adult male and female donkey, but when the jenny gave birth she was understandably quite protective of her newborn, especially against what she knew to be a predatory carnivore (even if a small one), and chased the cat away from her foal. Over the next few weeks, we observed the cat carefully work her way into the donkey family dynamic, hanging out mostly with the male and using every opportunity to play with the foal. Over a period of two to three weeks she was reintegrated into the donkey society, and was obviously accepted by the foal as a member of her social group. The clincher, however, was when the cat began to bestow "gifts" consisting of uneaten portions of small birds and mammals, placing these "offerings" carefully in the food bowls used by the donkeys. One has not truly lived until one has observed the reaction of a 300 kilogram jenny to the presence of vole hindquarters in her food bowl!

This curious social bond went both ways. One day Pierotti heard a ruckus, with all three donkeys braying. When he went to investigate, he found that "their" cat had trapped an intruder cat in a corner next to our chicken coop, where it was standing her off. Behind her in a semicircle stood all three donkeys, braying and showing aggressive threats toward the "usurper" cat. Pierotti got heavy gloves and caught the usurper (he later released it in an abandoned farmhouse several miles away). As soon as Pierotti put the stranger cat in a carrier, our cat and all three donkeys calmed down and wandered off together. Thus, Pierotti believes that his role in this mixed-species group was best described as "resolver of temporary crises" rather than "master and owner."

The reason we open with this story is to raise the question: into what category would you put the resident barn cat? She was the result of a domestic lineage, the domestic version of *Felis sylvestris,* but she lived almost exclusively on her own, or with other nonhumans, which made her at least a *feral* cat. In contrast, consider this definition of a *domestic* animal provided by an important scholar in the field of animal domestication: "one that has been bred in captivity for purposes of economic profit to a human community that maintains complete mastery over its breeding, organization of territory and food supply" (J. Clutton-Brock 1981, 21). No criteria in Clutton-Brock's definition apply to our barn cat, although this characterization could be applied to humans from Africa, stolen from their lands and peoples and transported to the United States and various Caribbean islands. Also interesting is that Clutton-Brock's definition clearly does not apply to the relationship between humans and wolves that we have described in the previous chapters. Although our barn cat was friendly with humans, even unfamiliar ones, she never entered a human home in her eighteen-plus years of life, even when offered the opportunity. In addition, how would you characterize her relationship with the donkeys, with whom she clearly had a strong social relationship? Were they her domestic animals? Her commensals? Or were they simply her social partners in yet another form of interspecies cooperation?

The cat's relationship with the donkeys and with humans clearly fits within Morey's evolutionary definition of "domestication": "Domestication . . . is best regarded as the development of a symbiotic ecological relationship between two organisms, with evolutionary benefits to both" (2010, 67). This definition is the most inclusive one we have found because it was designed specifically to include relationships that do not involve humans, such as that between leaf-cutter ants and fungi in the tropics. In addition, this definition clearly applies to the early stages of the relationship between humans and wolves.

According to the cooperative framework we propose, this early association fits within our understanding of social relationships; however, it would not fit within human-centered definitions of domestication (see Crockford 2006 and Morey 2010 for insightful discussions of this issue). Morey clearly follows the argument that "domestication was and is evolution" (Rindos 1984, 1). Crockford makes an emphatic point: "It's my firm belief that it was the animals that started it all, not us, and that the process must have been extraordinarily fast" (2006, 43). Both of these insightful archaeologists recognize that domestication is not a process conducted only by humans, and that it need not take a long time.

There are three basic terms, *wild*, *domestic*, and *tame*, used by most investigators when discussing human-nonhuman interactions. A fourth term, *socialized*, might be more appropriate than *tame*, which seems to be a catchall designation that has lost much real meaning in the contemporary world (Ritvo 2010). A fifth term, *feral*, is used to describe animals that have gone through the domestication process but return to living in a wild (or nontamed) state, as exemplified by the cat we describe above. One issue is that *domestic* and *tame* are often used interchangeably (Crockford 2006), although they have very different meanings. Complicating this issue is that *wild* is often used as if this concept were the opposite of both domestic and tame because of the tendency to conflate the latter two terms. What is often misunderstood is that it is perfectly possible for a nonhuman animal to be a domestic form but to be feral, not tame. Equally important, it is possible for an animal to be wild (nondomestic in all ways), but to be well socialized to live closely with humans. This is particularly true for the two species of carnivore with which humans have established long-term "domestic" relationships in the present world, *Canis lupus* and *Felis sylvestris*—dogs and cats.

Evolutionary archaeologist Susan Crockford points out that almost all definitions of *domestication* either state explicitly or imply that "humans made a conscious choice to tame certain wild animals for their own use or benefit, requiring that some animals be removed from the wild to form the domestic population" (2006, 34). She also emphasizes that domestication is a topic typically studied by anthropologists (including archaeologists) rather than by evolutionary biologists, even though Darwin (1859) used domestication as a model for his idea of evolution through selection. Domination of this concept by nonbiologists means that domestication has been insufficiently studied or even removed from consideration as an important evolutionary process, despite the statements made by Morey and Rindos above. Crockford argues that anthropologists cannot seem to agree on a definition of domestication, or even what precisely a domestic animal is, stating that when viewed solely from an anthropological perspective, "domestication . . . is always about us, what we did with the animals to bring them within our sphere of control. The animals were but hapless pawns in a well-orchestrated game we devised" (2006, 37). This way of thinking is readily shown by Safina, who states that "'domesticated' means genetically changed from wild ancestors by selective breeding" (2015, 221). Safina has a PhD in biology, but his definition is so narrow and human focused that it cannot be applied to domesticated reindeer, camels, elephants, or water buffalo, and we have no idea how dingoes might be classified according to Safina.

This lack of a solid definition, combined with the absence of evolutionary thinking and a focus only on human actions, is particularly important when considering the relations between *Homo sapiens* and *Canis lupus*. In the current anthropological framework, there is no room for individuals such as the Alaskan black wolf called Romeo (Jans 2015; see our conclusion to this book). Romeo was clearly socialized (or at least socializeable) to interact with humans, although he was clearly a wild wolf, neither domesticated nor tame, even if he behaved in a manner that suggested tameness. Such wolves may be much more common than is generally recognized—a number of individuals have recounted stories of wolves approaching them, soliciting play and generally acting interested in human companionship (Woolpy and Ginsburg 1967; Derr 2011). The authors themselves have observed instances of this behavior (discussed in chapter 11). In contrast, the domesticated canids seen on Cesar Millan's television show *Dog Whisperer* and described in *Cesar's Way* (Millan and Peltier 2006) are domestic animals in the classic sense. They may be tame, at least in a general sense, but they are often not well socialized (their owners have difficulty controlling them), which is a major reason why so many dogs are abandoned or even euthanized in contemporary America because of "behavioral problems."

Issues with socialization in supposedly domestic or tame canids can be seen in works as diverse as the bizarre, depressing anti-wolf screed *Part Wild* (Terrill 2011) and scholars' observations of border collies who are well tamed but poorly socialized, generating behavior that appears considerably wild (McCaig 1991). Pierotti recently learned more about these potential issues, as he is raising a cattle-herding border collie puppy, who is admirably socialized to his family but not to any other humans. She lives with an older Walker hound and a long-haired border collie, and these two dominant females have socialized her to canid society, enforcing very strict behavioral rules. Seeing how this intelligent, energetic, and independent young creature negotiates her social environment and life on a farm away from regular contact with nonfamily (or nonpack) members is a constant reminder that simply being domesticated does not always imply tame.

Domestication of nonhuman animals appears to be one of the least understood processes in evolutionary biology. It is rarely recognized that the importance of domestication in studying evolution represented one major distinction in the thinking of Charles Darwin and Alfred Russell Wallace, whose theories of evolution through natural selection were otherwise quite similar. Darwin recognized that processes involved in domestication clearly were related to evolution and selection, whereas Wallace argued strongly that the artificial human-driven component of this process made it irrelevant to evolutionary biology (Browne 2003). The debate continues to this day.

Ironically, Wallace's line of thought seems to prevail among anthropologists. Even scholars as insightful as Morey and Crockford stumble on this point because they do not seem to acknowledge that species such as

Homo sapiens and *Canis lupus* could have coexisted as ecological equals for thousands of years before the physical changes associated with domestication arose. As an example, in Morey's exemplary work *Dogs: Domestication and Development of a Social Bond* (2010, 75), he cites Crockford's argument (2006, 47) about "animals starting it all," and then spends several pages speculating about how wolf pups could have been introduced into human society. This, of course, assumes that humans initiated everything, because it is highly unlikely that wolf pups introduced themselves into human groups, especially before they were old enough to leave their natal dens. Lest readers think we exaggerate, we should point out that Morey introduces the topic with a discussion of the critical timing for the establishment of social bonds in dogs and wolves, which indicates clearly that he is discussing pre-weaned pups. A far more logical assumption is that these pups were still part of their original pack, and it was the pack, or at least a gravid female, that affiliated with humans. There are also individual animals like Romeo who, even though he was a young adult male, was clearly willing to associate with humans and their domestic canids, forming strong bonds with at least one human and his dog (Jans 2015).

Some of this conceptual conflict arises because although they recognize the possibility of earlier affiliations between humans and wolves, neither Morey nor Crockford is comfortable pushing this date back past the 15,000 YBP barrier, at which time phenotypically smaller canid specimens start showing up in the archaeological record. When their works were originally published, no unequivocal dog specimens were known that were older than 14,000 years old, a situation now changed in the intervening years since the 2010 publication of Morey's book. In any case, Morey's concerns about what wolf pups would eat or how they would learn hunting skills are negated if we assume that entire packs, or at least females and pups, affiliated with humans, and everyone learned together. Shipman (2015) assumes a middle ground: she makes statements about stealing pups while also referring to groups of females with regard to the findings of Germonpré et al. 2009; Germonpré, Lázničková-Galetová, and Sablin 2012; and Germonpré et al. 2013, 2015. This situation would also negate an idea put forth by Serpell, who argued that the route to domestication of dogs must have involved "killing or driving away temperamentally unsuitable individuals" (1995, 104). If wolves were free to come and go of their own volition, as we describe for dingoes in chapter

6 and for wolves in chapter 5, such a situation would not necessarily arise. Morey further argues that wolves "scaveng[ed] human kills" (Morey 1990, cited in Morey 2010, 69), and presents a sort of quasi-endorsement of the Coppingers' (2001) "refuse dump" idea, which he also critiques because it is irreconcilable with his own expertise on human/dog burial sites.

The key point here is that these scholars did not engage with stories and accounts from Indigenous peoples, which undermine much of their otherwise insightful arguments. Rather than a scenario of wolves scavenging human kill sites, accounts from several Indigenous societies indicate essentially the opposite, contending that humans might not have survived without being able to scavenge or even steal kills made by wolves (Schlesier 1987; Marshall 1995; Pierotti 2011a). Acknowledging Indigenous peoples' accounts of wolves providing food for humans could have spared these otherwise solid scholars some unfortunate assumptions about wolves coming begging to humans rather than the other way around, which would change the entire dynamic of the relationship between the species. It would also reveal why many of these peoples identified wolves as creator figures responsible for teaching humans to hunt, a concept completely missing from any anthropological accounts of human/wolf dynamics but present in discussions of Indigenous cultures (see Schlesier 1987; Marshall 1995; Pierotti 2011a).

A general lack of understanding domestication as an evolutionary process seems to result because of European or Euro-American cultural biases. As indicated in Darwin's epigraph at the opening of this chapter, the classic assumption is that breeders select for specific desired features and often remain largely ignorant of the genetic basis that underlies such features and the actions of selection. It is typically assumed that domestication results only from human action and involves conscious decisions about the type of animal desired by humans (e.g., Safina 2015) rather than from a coevolutionary relationship that shapes both species over thousands of years, as we argue in this work. Crockford describes how "great success over recent centuries in manipulating animal breeds, and creating new ones" led to the assumption that early transformation of animals must have involved conscious decisions by humans, and that such dramatic cultural change leading to "progress in ancient human societies had to be both intentional and premeditated" (2006, 39). As pointed out in chapter 4, this assumption also underlies the debate between Germonpré and her colleagues (Germonpré et al. 2013, 2015) and critics of their findings (Morey 2010, 2014; Morey and Jeger 2015). Crockford argues that this human-centered view of domestication is an idea constructed entirely from assumptions, unsupported by any facts and functioning largely as "widely accepted myth" or dogma, contending, "Biologists are as accepting of this idea as the anthropologists who promoted it in the first place" (2006, 40), but she finds it heartening that some scholars seem "willing to abandon the dogma and acknowledge that some wild animals may simply have chosen to live [closely with humans] and changed in consequence" (42).

Crockford is trying to answer the question "If people didn't deliberately create domestic animals, how did they come about?" This is an important question, one somewhat convergent upon the approach we take in this work; however, Crockford believes this question is linked to the issue of "how any species of animal can transform into another" (2006, 42). This line of argument suggests that Crockford assumes that wolves and dogs are different species, an issue we have addressed at several points. The key to understanding the relationship between dogs and wolves is that, despite phenotypic differences in some cases, they all remain members of *Canis lupus* (Pierotti 2012a, 2012b). Crockford's insightful argument is based on the assumption that changes in amounts of thyroid hormone produced in different forms may be responsible for the juvenilization of wolves. We suspect that she is right in identifying thyroid regulation as a factor affecting phenotypic change, especially neoteny, which involves animals speeding up their growth rate so that they become physiologically mature without becoming physically mature. Such regulatory changes are much more likely to be *epigenetic*—changes in gene regulation rather than in the genes themselves. This means that such changes are unlikely to lead to speciation but could very well lead to the production of new, unusual phenotypes within short time periods. Crockford (2006) seems to be in basic agreement with our argument that the changes involved in domestication could occur within a single century or less (see also G. K. Smith 1978).

Actual speciation events require significant changes in the genes themselves (as opposed to epigenetic "switches"). This is the only way that long-term evolutionary change can take place, and it probably takes longer than the short time periods Crockford invokes. From a population genetics perspective, the actual consequences and process of selection, along with the associated epistatic (multiple genes affecting one trait) and pleiotropic (one gene affecting multiple traits) interactions, go largely ignored in studies of domestication. Human-driven selection in many domestic dogs is directed at nonadaptive, or even maladaptive, anatomical features such as short muzzles, fluffy or frizzy coats, or even hairlessness. In other species such as cows, pigs, or chickens, life history features are selected for, such as milk or egg production, and these, like most life history traits, almost certainly have low heritability (Roff 1992). Behavioral characters such as ability to be handled by humans without showing resistance, biting, or struggling would also have low heritability (Hemmer 1990).

One problem with understanding the selection process involved in domestication and how it affects specific character states is that there appear to be pleiotropic interactions between behavioral and anatomical traits (possibly physiological and life history traits as well), which confound the standard assumptions of how selection acts upon a specific trait. Pleiotropy means that the same gene—or, more realistically, the same gene complex—has effects on several traits displayed by a particular organism. For example, in the classic studies of Belyaev and Trut in Siberian fox farms (see below), selection for "tameness" ended up also favoring correlated traits such as changes in coat color, ear configuration, leg length, skull shape, dentition, and so on (Trut 1991, 1999, 2001; Morey 1994, 2010; Trut et al. 2000; Crockford 2006). Most domestic selection on behavior seems to act primarily through changing rates of development (ontogeny), especially in selecting for the retention of juvenile characters into adulthood (referred to by evolutionary biologists as *neoteny*). Juveniles are easier to handle than adults; therefore, to most people, including scientists, the tameness found in newborn or juvenile individuals also generates neotenic changes, manifested through physical as well as behavioral traits.

From our perspective, domestication is an evolutionary *process* rather than a definable state or endpoint, even though it rarely, if ever, leads to speciation (contra Crockford 2006). One perceptive student of domestication has stated, "Domestication was and is evolution" (Rindos 1984, 3). In this overall context, we echo Phillipe Descola's question of "whether the opposition between wild and domesticated can have been at all meaningful in the period prior to the Neolithic tradition, i.e., during the greater part of human history" (2013, 33). This is pertinent especially because the Indigenous peoples we discuss were living in preagricultural societies.

Evolution is a process rather than an outcome or set of outcomes (Pierotti 2011a). At many times in the past, humans established relationships with wild species that may have led to a state of domestication. As another scholar puts this: "Wild and the tamed or domesticated exist along a continuum. . . . Few animals live lives untouched by [humans]. . . . In this sense, few can be said to be completely wild [e.g., wolves in Yellowstone were acclimated in pens prior to release]. As the valence of the wild has increased and its definition has become assertion, rather than description, the boundaries of domestication have also blurred. *Not that they were ever especially clear.* . . . 21st century wolves belong to a long line of animals where 'wildness' has been compromised, tameness has also existed on a sliding scale. . . . Both 'wild' and 'tame' have existed for a millennium, remaining constant in form as well as in core meaning, while the language around them has mutated, beyond easy comprehension, if not beyond recognition" (Ritvo 2010, 208).

Human relationships with the two species of carnivore currently domesticated by humans—dogs and cats—are clearly different than relationships humans have established with domesticated ungulates because the carnivores were cooperating with humans rather than being raised for food. Although dogs and cats are clearly domesticated today, in both cases the relationship almost certainly began as commensalism, in which individual wild wolves and wildcats established close, but highly variable, social dynamics with individual humans. In wolves, this association was almost certainly linked to hunting activities, as described by numerous Indigenous peoples (Schlesier 1987; Corbett 1995; Fogg, Howe, and Pierotti 2015; Shipman 2015). In contrast, cats seem to have figured out that once humans established large-scale agriculture, storage granaries attracted large numbers of birds and rodents. This represented easily acquired prey, especially for pregnant or lactating females. Kittens produced by these females grew up familiar with, and presumably unafraid of, humans. Humans quickly learned that cats reduced "pests" around their agricultural activities and thus encouraged their presence. At this point the relationship would be clearly commensal. Such relationships initially emerged from a commensal or coevolutionary ecological relationship and persisted for thousands of years in this form. Only in

relatively recent times has there been obvious selection of specific phenotypes and deliberate breeding by humans, contra Safina (2015, 221).

With reference to human-driven changes, Hemmer 1990 describes
three principles of domestication: (1) the species must be able to breed
under captive conditions, (2) individuals with smaller relative brain sizes
are more prone to domestication, and (3) selection for appearance (e.g.,
coat color) can elicit domestication effects or vice versa (selection for
friendly behavior might elicit changes in appearance) (Belyaev 1979;
Belyaev and Trut 1982; Morey 1994, 2010; Shipman 2015). After only a few
generations of living with and breeding under the influence of humans,
carnivores can be well on the way toward domestication, depending upon
the strength of the selection imposed by humans (G. K. Smith 1978;
Belyaev and Trut 1982; Crockford 2006; Pierotti 2012b); however, this
process can readily reverse. Even today, many cats retain the ability to
switch between a domestic and feral state in which they live without
depending upon humans (Dombrosky and Wolverton 2014), as Pierotti
and many people familiar with cats can attest from experience.

Among commensal social species, individuals can return to living
without humans, especially during early stages of domestication, whereas
others may remain with humans for many generations and still retain the
wild phenotype if there is no selective pressure to generate a different
phenotype, such as in Asian elephants (*Elphas indicus*) or dingoes. Some
ungulate species remain in a slightly altered state, functioning as quasi-
or proto-domestic (Crockford 2006)—for example, reindeer (*Rangifer
tarandus*), yak (*Bos grunniens*), water buffalo (*Bubalus bubalus*), and drom-
edary and Bactrian camels (*Camelus dromedaries* and *C. bactrianus*)
(Hemmer 1990).

Even the most extreme domestically generated phenotypes probably
exist as potential phenotypes present within the genotype of the nondomestic ancestors. Miniature dogs or mastiffs remain wolves in their genes
and their internal structure (cells, tissues, and organs); however, these are
recent breeds, no more than a few centuries old, and such forms would not
have survived during coevolutionary stage s of domestication. Various types
of wolves underwent differential selection as a result of their interactions
with humans, and this process eventually led to phenotypes that we now
consider to be domesticated. In recent centuries, this variety, combined with
human-driven selection, resulted in domestic dogs becoming the most

anatomically diverse mammals within a single taxonomic unit on the planet, even though their wild ancestors seem to differ primarily in size and coat color (Morey 1994; Vilà et al. 1997; Spady and Ostrander 2008; E. Russell 2011; Hunn 2013). Descendants of these different events almost certainly encountered one another over the millennia and crossed repeatedly, further rendering attempts to tease out DNA relationships difficult, if not impossible (Coppinger, Spector, and Miller 2010; Pierotti 2012a, 2012b, 2014). If we include deliberate breeding of wolves back into various "dog" lineages (G. K. Smith 1978), the problem becomes even more intractable.

SIBERIAN FOX STUDY

In 1959 at a fur farm in Siberia where silver foxes (*Vulpes vulpes*) were being raised for their pelts, a long-running experiment was initiated with the goal of studying the behavioral genetics of domestication (Belyaev 1979). Investigators noticed that while most individual foxes were quite mistrustful of people, about 10 percent seemed to be less fearful. To examine the heritable basis of this less fearful (more friendly) behavior, a strict selective breeding program was established in which calmer individuals were selectively crossed only among themselves (Belyaev 1979; Belyaev and Trut 1982). The experimental population was chosen based on a single criterion— "tameness" or "friendliness," determined by whether foxes fearlessly and nonaggressively approached a human. A second population of foxes exhibiting the more common "fearful" behavior was maintained as a control. This latter population was bred randomly in respect to their behavioral responses toward humans (Trut 1999, 2001).

After twenty generations, foxes in the experimental group showed some remarkably "doglike" traits, both physical and behavioral. "Tame females exhibit statistically significant changes in certain neurochemical characteristics in . . . the hypothalamus, midbrain, and hippocampus" (Belyaev 1979, 306). Selection based solely on the criterion of tameness led to behavioral, physiological, and morphological changes in the experimental fox population that were not observed in the control population. Foxes in the selected group showed little fear and aggression toward humans, instead demonstrating affection and soliciting attention from people. This reduction in aggressiveness and fear appeared to result from attenuated activity of the pituitary-adrenal axis (Plyusnina et al. 1991), lending support to Crockford's (2006) thyroid hormone idea.

Alteration of serotonin, noradrenaline, and dopamine transmitter systems in specific brain regions linked to regulation of emotional-defensive responses was found in the experimental foxes (Trut 1991, 2001). After more than forty generations, it appeared that the behavioral changes and associated phenotypic physical traits were clearly heritable (Trut 1999, 164–65). These changes appear to be consequences of changing allometric patterns within development because experimental foxes show not only behavior changes but also higher frequencies of floppy ears, short or curly tails, nonpigmented hair, extended and early reproductive seasons, and changes in the size and shape of the crania and dentition (Trut 1991, 2001). Similar traits are also found in various breeds of domestic dog, which suggested to some that selection for "tameness" in wolves favored retention of juvenile traits, such as short legs and floppy ears into adulthood (neoteny). A check of farm breeding records revealed that females in the experimental, less fearful population that had given birth prior to initiation of the experiment were all early breeders, ovulating in the first ten days of the six to eight week mating season (Crockford 2006). Early breeding is typically associated with superior phenotypes in many species (Pierotti 1982; Annett and Pierotti 1999). In this case it led to females who came into heat more than once within a twelve-month cycle, another trait found in many domestic dogs but not in Russian laiki, Saarloos wolf-dogs, and some other large wolflike breeds.

Although we find this fox research program very interesting, it may not be as significant in allowing us to understand the early stages of the human-wolf relationship as is commonly supposed (e.g., Morey 1994, 2010; Crockford 2006; Shipman 2011, 2015), largely because the physical changes observed in the experimental fox population have not been found in stories from Indigenous peoples or in archeological specimens from the first 20,000-odd years of the human-wolf relationship. The Russian study involved intense selection on a presumed single trait—sociability or "tameness"—which actually turned out to be a complex of traits involving life history, physiology, anatomy, and behavior. In practice this selection seems to have led primarily to rapid growth and development rates, which are known from life history studies to lead to early reproduction (Roff 1992) and also seem linked to neoteny. As suggested by accounts from Indigenous peoples described earlier, during early stages of human-wolf relations, humans were quite happy with original nondomestic wolf phenotypes and did not want overgrown puppies as companions. Many wolves seem to be easily socialized to live with humans without showing physical changes in coat color or obvious changes in physical traits (G. K. Smith 1978; Jans 2015), as Pierotti has observed. This suggests that wolves can be socialized without quickly developing neotenic traits.

Only two findings from the Belyaev/Trut fox work seem to apply to the domestication of wolves by early humans. First, selection that is directed at tameness or sociability has been shown to lead to associated morphological changes within short periods of time, such as those found in the "wolves" identified as early dogs (Germonpré et al. 2009; Morey 2010; Ovodov 2011; Germonpré, Lázničková-Galetová, and Sablin 2012; Shipman 2015). Such observed differences were used to identify the "first dogs" (Shipman 2015), and may be the result of mild selection imposed by changes in environmental conditions—that is, animals choosing to remain with humans and interbreeding—rather than the consequence of selective breeding by humans. Second, the results of the fox study suggest that it is possible to rapidly produce animals with suites of neotenic features found in the ancestors of many contemporary dog "breeds." This supports arguments made by Crockford (2006) and should effectively silence the oft-stated nonsense that it took 10,000, 12,000, or even 30,000-plus years after humans started selectively breeding wolves or wolflike dogs to produce contemporary dogs from wolves (see chapter 10 and Pierotti 2012a). Behavioral changes related to sociability but not obvious morphological changes can be easily achieved within five to ten generations of strong selective breeding, even starting from pure wolves (G. K. Smith 1978).

WHY DOMESTIC FORMS ARE NOT TRUE SPECIES

The issue of whether domestic forms should be recognized as distinct species is especially problematic in the transformation of wolves (*Canis lupus*) into domestic dogs (figure 9.1). The result of the outdated practice of being given a Latin binomial, *Canis familiaris,* by Linnaeus (1792) in the eighteenth century has been that many otherwise respectable scientists make the twinned mistaken assumptions that domestic dogs represent (1) a single evolutionary event and are thus (2) a valid species. These assumptions are linked to domestication dogma described by Crockford (2006): that is, changes in thyroid hormone related to self-imposed

FIGURE 9.1   Wild *Canis lupus.* Note the lowered head, wide-apart ears, relatively short snout, hind legs at ninety-degree angle to hips, and tail not reaching past hocks. (Wikipedia)

selection for sociability could lead to formation of new species. Although these arguments might be valid for generating phenotypic changes, the revolution in DNA sequencing over the last two decades has made it apparent that *domestic* forms of all kinds show a range of genetic attributes that suggest they are not separate species from their ancestral forms but variant phenotypes lurking within the genotypes of their *wild* ancestors.

Contemporary domestic dogs are a grab-bag assemblage from which certain lines have arisen at different times and in different places over the last 30,000 years or so (Morey 1994; Derr 2011; Shipman 2011). As a result, the assemblage known today as "dogs" appears to be not only *polytypic* (containing multiple body and behavioral types or *phenotypes*), but also *polyphyletic* (almost certain to have multiple origins) (Morey 1994; Coppinger, Spector, and Miller 2010; vonHoldt et al. 2010; Pierotti 2014; Frantz et al. 2016). True species cannot be polyphyletic because to be a species a lineage must have a single origin (Pierotti 2012b, 2014). Contemporary domestic dogs include a wide range of phenotypes, ranging from some recognized breeds almost indistinguishable from their wolf ancestors through intermediate forms to dwarf phenotypes that at best resemble mutant forms of wolf puppies, such as Chihuahuas, cocker spaniels, bichon frises, Yorkshire terriers, Pekingese, and pugs.

Despite their anatomical and behavioral oddities, all of these breeds continue to carry wolf genes and to show wolflike physiological and life history characteristics, such as number of offspring per litter, length of gestation, state of development at birth, and so on. Domestic dogs are the most common form of household pet in the Western world, with numbers worldwide being estimated at around half a billion (Coren 2012, 228). For breeds that are superficially most like wolves in their appearance, that is, their external phenotype, there is often confusion, especially in Europe and North America, about whether or not a given animal is a wolf, "wolf-dog," or simply a "dog" (see chapter 10). This is especially true when breeders cross wolflike breeds or cross wolves with various breeds of wolf-like dog. We suspect that most people, including almost all individuals involved in animal control (or dog catching), would identify a cross between an Alaskan malamute and a German shepherd (figure 9.2) as a wolf-dog because both species resemble wolves in their basic phenotypes, and the cross looks more wolflike than does either parental breed. The wide-apart ears of the malamute balance the closer together ears of the shepherd. The continuous coat color of the shepherd balances the dramatic boundaries between black and white in color of malamutes. Perhaps most important, the upright stance of the malamute negates the AKC preference for German shepherds with malformed hips that cause their legs to extend well back behind the torso and often lead to hip dysplasia.

The tendency to try to distinguish between wolves and dogs in a consistent way creates some logical errors. As an example, geneticists who study canids suggest that German shepherds have no more wolf ancestry than other "dogs" (e.g., Vilà et al. 1997; vonHoldt et al. 2010; but see Frantz et al. 2016). The history of the breed suggests otherwise. German shepherds (Alsatians) did not really appear as a distinct pheno-type until the 1890s. The original studbook, *Zuchtbuch für Deutsche Schäferhunde* (SZ), lists four wolf crosses, from SZ no. 41 to SZ no. 76 (von Stephanitz 1921). Thus, among the animals used in the development of the breed of dog known in America as the German shepherd, at least four were descended from dogs with recent wolf ancestry. European Alsatians are much more wolflike in appearance than their American

FIGURE 9.2 This cross between an Alaskan malamute and a German shepherd superficially resembles a wolf more than either parental breed. It is crucial to understand that this animal has no recent wolf ancestry. The nearest wolf ancestors would be the wolves that gave rise to the respective breeds at some time in the past. (Photo provided by owner, who wishes to remain anonymous)

counterparts, which may be responsible for some of the bias (see below). The originator of the Alsatian breed wanted a much more wolflike pheno-type to avoid the trait of hind legs canted backward that ruins American dogs. In 1899, Max von Stephanitz, attending a dog show, was shown a dog named Hektor Linksrhein who was allegedly one-quarter wolf (figure 9.3). The dog, renamed Horand von Grafrath, and his progeny were used to create the Alsatian wolf-dog, known in America as the German shepherd. Von Stephanitz wrote a book about his new breed, *The German Shepherd in Word and Picture* (1921). He documented the heritage and pleaded to breeders not "to add more wolf blood" into his dogs as he had already found the ideal combination. This single male became the core of the breeding program and was bred with dogs belonging to other society members that displayed desirable traits. Thus, there is docu-mented use of pure wolf genes to create a domestic dog breed, the German shepherd, which has been characterized by geneticists as being completely distinct from wolves (vonHoldt et al. 2010). The wolf-dog used by von Stephanitz was the male parent; therefore its influence would not show up in the mtDNA used for at least some of the results reported by

FIGURE 9.3 Horand von Grafrath, the first German shepherd dog (1895–99). This individual, allegedly 25 percent wolf, was used to found the breed. The animal is much more wolflike than the popular contemporary breed. (http://retrieverman.net/2010/10/21/ do-german-shepherds-have-wolf-ancestry/)

geneticists because mtDNA is inherited only from the maternal line. American wolf breeder Gordon Smith (1978) states that some of his domesticated wolf lines had their papers switched with AKC-registered American German shepherds in order to improve the line.

DOMESTICATION IN *CANIS LUPUS*

Understanding cultural and intellectual traditions is crucial to understanding what different people mean when they use the term *domestic* or when they try to define when an animal (or type of animal) becomes domestic. Ray and Lorna Coppinger argue, "There is zero evidence for dogs before 12,000 BP" (2001, 286) and assume that it took all the ensuing time to create the *modern dog*, although they are unclear on what this actually means. Their idea of *dog* seems to be the pariah dogs found in much of the world's tropics and subtropics—jackal-sized, scrawny, short-haired animals that are rarely cared for by humans. Although such creatures are interesting because of their feral nature, where they must forage for themselves, these animals are not wolflike, nor are they similar to the first wolves who lived with humans. The unexplored problem, however, is that scholars are rarely specific about what they mean when they use the term *dog* (see Hunn 2013), while at the same time it is clear that few of them have in mind the kind of animals that look like the first wolves to have associated with humans. This dissembling and confusion can have serious public health and legal ramifications (see below).

A key issue we raise throughout this book is the distinction between behavioral changes and anatomical features (physiognomy). Many American and European dog owners want animals that closely resemble wolves in their appearance but *behave* like dogs, well socialized to humans (see G. K. Smith 1978). This issue is problematic because, as any regular watcher of Cesar Millan's popular television show *Dog Whisperer* will recognize, many obviously domestic dogs are not well socialized and can be aggressive toward humans. This is the dark secret at the heart of the relationship between humans and wolves, both wild and domestic: canids are carnivores that naturally live in social groups, but the dynamics of those social groups are characterized by a mixture of peaceful cooperation and aggression, both implicit and explicit (Spotte 2012). If you coexist with a group-living carnivore, it will strive to figure out where it fits into the social dynamic of your household. This will involve testing the limits of what it can do within this dynamic, which creates behavioral conflicts.

As an example, at the age of five weeks, one of Pierotti's wolf/German shepherd cross puppies encountered his wife, Dr. Cynthia Annett, who had been working out of town when the puppies joined the household. This animal, the dominant puppy in her litter, approached Dr. Annett growling, her hackles raised, her skinny little tail sticking out behind her like a flagpole. Dr. Annett, an expert on social behavior and aggression in nonhumans, grabbed her by the scruff of the neck, flipped her on her back, and said, "I'm the head bitch here, don't ever forget it." From that moment forward the two of them had an amiable relationship, with no aggressive interactions, which lasted until the death of the animal from natural causes eleven years later.

This raises an interesting issue concerning the nature of dominance within wolf packs and in human/canid relationships. Excessive reliance on this concept has led to many incorrect assumptions because it emerges from an understanding of nature and its workings as hierarchical. To many people, belief in this concept suggests that humans must constantly show dominance over both their human and their nonhuman companions (Millan and Peltier 2006; Spotte 2012). This usually means that any time a puppy, or even a grown dog, shows any hesitation or resistance to a command, the appropriate response is to body slam the animal to the ground in what is described as an "alpha roll." In contrast, David Mech, an expert on the behavior of wolves in the wild, has demonstrated that dominance hierarchies in wolf packs are largely artifacts of captivity, and that he rarely if ever observed contests over dominance in his many years of research on free-living wolves (Mech 1999; Spotte 2012). Pierotti's personal experience with wolf-dogs and wolves supports this conclusion, because once a human clearly shows that he or she is in control, the animals relax into accepting this dynamic. Cesar Millan constantly emphasizes the point that you want your dog in a state he refers to as "calm, submissive," by which he seems to mean that the animal is relaxed and accepts the leadership of the human (Millan and Peltier 2006). To Mech, dominance is basically the same as parenting. As in a human family, an effective parent does not need to constantly play out dominance conflicts with his or her offspring. In the human relationship with a puppy, the young animal is simply seeking consistent knowledge of what is acceptable and unacceptable behavior on its part. Thus, Dr. Annett's behavior toward a five-week old wolf-dog cross might be seen by some as dominance, but it is better interpreted, as she put it, as the "head bitch" showing a subordinate what was unacceptable. At times it may become necessary to put an animal into a submissive position, but these occasions should be rare, resorted to only when clear violations of social protocol are involved. This was a mistake made by the author of *Part Wild* (Terrill 2011); the human failed to establish any effective rules with a large, friendly, but rambunctious dog, leading to a tragic and troubling outcome (see chapter 10).

We prefer the term *socialized* rather than *tame* because it is obvious in many cases that the relationships that emerge between humans and their nonhuman companions are a result of both species learning to accommodate the behavioral characteristics of the other. *Taming* implies that the human is automatically dominant within this relationship, but the reality is much more nuanced and complex. It has become fashionable to criticize Cesar Millan for using the supposedly refuted "dominance technique"; however, observation of him at work shows that what he does is quickly and effectively establish rules, and then try to teach the humans in the relationship how to replicate this behavior. As Millan points out on nearly every episode, he doesn't train dogs, he trains people. Most problems with dogs, and wolves or wolf-dogs, result from three basic causes: (1) humans allow the animal to become anxious because relationships within the social group are not clearly defined, (2) the animal thinks it is dominant over, or at least has an unresolved relationship with, one or more human member of the social group, or (3) the humans have allowed the animal to become territorial in regard to either themselves or another family member (the third issue emerges from the second).

In contrast to *socialized* or *tame,* where wolves are concerned, *domestic* implies that humans have been specifically selecting for certain traits that they find attractive or valuable. Most people assume that domestication involves changing the appearance of an animal, but as in the Siberian fox study, this process can involve changing behavioral traits as much or more than physical ones (Hemmer 1990; Trut 2001; Crockford 2006; Spotte 2012). In the case of the domestication of wolves, it is likely that, contra to the model proposed by the Coppingers (2001) and some arguments by archaeologists, behavior in wolves changed well before any physical traits did. This is also the resolution to the debate we discussed in chapter 4. People today who are trying to produce a wolfy-appearing dog are more interested in changing behavior without obviously changing appearance, a situation that appears to have happened in Siberian laiki.

DEFINING WOLF, DOG, AND WOLF-DOG

Pierotti has served as an expert witness of wolf and dog identification in eighteen cases from 1993 through 2003. We draw on those experiences to discuss how people perceive the process of domestication in wolves. A miner in Alberta, Canada, owned an animal that was identified as "pure wolf" by the provincial wolf biologist (figure 9.4), a person with a solid research record working with wolves in their natural environment. Once Pierotti encountered the animal in the flesh, however, it was easy to recognize several mistakes the wolf biologist had made. To begin with, the biologist made only a single brief (less than fifteen minutes total) visit to the animal and refused to enter the enclosure in which it was kept. This made sense from the perspective of a wolf biologist because to him and his cultural concepts, the animal was a "wolf" and should not be interacted with socially. Pierotti, on the other hand, sensing that the animal was well socialized, entered the enclosure, at which point the animal responded as

FIGURE 9.4 Based on external appearance alone, the
provincial wolf biologist of Alberta identified this animal as
"pure" wolf. Note the backward sloping legs and powerful
neck and shoulders, which are characteristic of some
AKC-registered breeds but not of wolves (cf. figure 9.1).
(R. Pierotti's personal collection)

if greeting a long-lost friend, romping and roughhousing and showing joy
and excitement, not shyness or fear. No pure wolf would act this way with
a strange human. Pierotti uses playful interactions with animals as opportunities both to assess behavioral tendencies and examine general physiognomy. In this case Pierotti quickly discovered that the dog had a
barrel-shaped chest and powerful front legs and shoulders, indicative of
malamute ancestry. The wolf biologist had focused on the outside and
saw "wolf"; Pierotti concentrated on the inside and quickly realized this
animal was not a wolf. Its bones, especially in the legs, were heavy. It also
had hind legs that sloped backward and a pink nose, both of which are
traits found in some dogs but never in wolves (figure 9.5).

The turning point in the case was when the defense attorney asked
the wolf biologist, "How familiar are you with dogs?" His response was,
"When I was younger, I had a dog." This revealed that from his cultural
and personal perspective, the animal did not fit his personal image of
"dog" and thus must be a "wolf," especially based on his superficial examination. This is an example of familiarity bias, where individuals see what
they are most familiar with regardless of what may actually be present.
Further questioning revealed that the wolf expert was unfamiliar with
Alaskan malamutes and could not describe the breed or name any breed
characteristics, which sums up the problems that arise in almost every

FIGURE 9.5 A face-on view of the Alberta "wolf"
reveals its pink nose, a classic malamute trait, as well
as its wide shoulders and chest. (R. Pierotti's personal
collection)

such case. Experts on "wolf identity" show a form of familiarity bias when
they fail to consider the doglike aspects of an animal and look only at what
they consider to be its wolflike characters, primarily the shape of the head
or the color of the pelage.

Much of this debate about the relationship between wolves and dogs
hinges on the definition of exactly what is a wolf or a wolf-dog. Most people
are incredibly ignorant on this point and think that any dog that looks
wolflike to them is a "wolf-dog." In cases where Pierotti has served as an
expert witness, well-recognized breeds such as Eskimo dogs, Siberian
huskies, Alaskan malamutes, Samoyeds, and even Belgian sheepdogs,
along with a selection of fairly obvious "mutts" have been identified as
"wolf-dogs," or even as "wolves" by animal control officers (chapter 10).

Further, most people are completely ignorant of breeds such as the
Saarloos wolfhound, Czech wolf-dog, or west Siberian laika, all of which

are well-recognized breeds in Europe. European and Russian breeders have created new breeds of dog by crossing wolves and German shepherds or aboriginal dogs and then crossing the offspring within recent times (see G. K. Smith 1978). Like wild wolves and unlike most domestic breeds and the experimental population of foxes described above, these breeds typically have only a single estrus period per year, usually in February–March (the wild wolf breeding season), and the first estrus occurs at an age of one and a half to two and a half years. Domestic dogs, in contrast, usually come into estrous during their first year, may come into "heat" as early as six months of age, and have multiple periods of estrus per year (Spotte 2012).

This opens up a set of interesting questions. Most important, what exactly is a "wolf-dog"? If people cross *wild-type* (natural) phenotype wolves to see if they can produce a tame and friendly animal that has a "doglike" personality, they are probably doing something similar to what early humans did to produce the earliest "dogs," albeit with a more obvious purpose in mind (see, e.g., G. K. Smith 1978). The next question is, are such animals dogs or simply wolves subject to selective breeding— that is, the domestication process? The irony is that, at least from the point of view of contemporary systematists, wolves and dogs are essentially the same, because all are *Canis lupus*. This is what we mean when we say, "All dogs are wolves, but not all wolves are dogs."

Another significant question arises after a couple of generations when the animals in such breeding programs are generally friendly and social with humans: are they "wolves," "wolf-dogs," or simply "dogs"? According to the definition provided by Clutton-Brock in the opening pages of this chapter, they are domesticated. This is where the key definitional issues lie, as well as the point upon which most of the cases on which Pierotti has served as an expert witness have pivoted. That is, will the judge be convinced that the animal whose life may be at stake in the proceedings is functionally a dog, not a "wolf-dog" or a "wolf"?

Pierotti is currently involved in a case in Massachusetts in which this issue is paramount. The owner obtained dogs from a breeder who specializes in producing animals that look like wolves but are completely socialized to humans, especially for films and television, where "wolves" are often required but actors do not want to interact with an unsocialized or poorly socialized animal. The animals in question look very much like wolves superficially; however, they are among the friendliest large dogs Pierotti has ever encountered. The male is a very large puppy, weighing over 100 pounds at nine months old, which is much too heavy for a wolf (see chapter 10). His nose is turning completely pink which, along with his size, reveals his malamute ancestry, just as in the Alberta case (above). The female is predominantly black with some scattered white hairs, and looks more like a classic wolf; however, she does not behave like a wolf in any way. Both animals have no pure wolf in their recent ancestry; they are products of at least ten generations of crossing. The Massachusetts wildlife authorities insist that these are wolves, and as in Alberta they refuse to interact with the animals, although one expert testified that the animals seemed "friendly," which he found surprising.

The key issue in the case seems to be: can authorities accept the idea that it is just as easy to breed selectively for "wolfy appearance" using dogs as it is to select for floppy ears, shaggy coats, heavy legs, or any other trait desired by humans? These animals are clearly domestic, and as domestic specimens of *Canis lupus,* they are by definition dogs. The authorities are contesting this, showing yet another example of obvious familiarity bias and lack of knowledge.

What creates situations like this is that many people claim to be able to identify "wolf-dogs"; however, these "experts" are rarely familiar with the range of potential breeds that can be used to produce an animal that appears superficially to be "wolflike," yet stubbornly cling to their identifications, sometimes even after they have been shown to be obviously wrong (see chapter 10). We mentioned above cases where experts had identified Belgian sheepdogs and Alaskan malamutes as "wolves." Some "experts" have identified as wolves animals so obviously non-wolflike that it is laughable, but it is possible that the court may take them seriously, especially if the defense has no expert to counter their testimony.

An example of this problem is when Cesar Millan, the "Dog Whisperer," devoted an entire episode to the issue of "wolf-dogs." We want to preface this by stating that we have great respect for Millan and his techniques (Millan and Peltier 2006), but in this show he essentially ceded narrative control to a young woman, identified as a "wolf expert," who, although charming, was either confused or dissembling. The episode focused upon three individual animals in three different households, only one of which Pierotti would have identified as having any appreciable amount of recent wolf ancestry. The first individual was pretty clearly a malamute/husky type dog, with a prominent stop (forehead) and a tail that it carried curled over its back. In any of the cases in which Pierotti has participated, these traits alone would have been sufficient to get the case thrown out of court. The animal was friendly and greeted strange humans with affection, which, as we discussed above, is doglike, not wolflike, behavior.

The second animal appeared to be somewhat "wolflike," in that its facial configuration looked right and it was very shy of unfamiliar humans. The "expert" did not discuss the obvious differences in behavior and physiognomy between these two animals, which should have been essential information, both for Millan and the viewing public. The third animal, whose owner blithely stated that she had had several previous "wolf-dogs," was an unusual dog that was suffering from extreme social anxiety when left alone. Like the first animal, this dog was very friendly and readily approached unfamiliar humans. This behavior was so obvious that even the "expert" commented that this was "very unusual for a wolf-dog."

Any large dog can be a problem if it is not well socialized or has been able to assume a dominant role within its family (pack) (Coren 2012; see http://www.dogsbite.org/dangerous-dogs.php for the data). The most threatening canid Pierotti has ever encountered personally was a German shepherd owned by a family friend. This animal could not be approached by unfamiliar people; in fact, it was aggressive to everyone in its human family but the mother and the younger son. This animal had much bigger issues than any of the "wolf-dogs" featured on *Dog Whisperer,* and its behavior and aggressive nature were in complete contrast to the "wolf-dogs" identified by the Massachusetts wildlife authorities, which are being prosecuted because the authorities regard them as "dangerous."

LEGAL AND PUBLIC HEALTH CONSEQUENCES OF THE CONTROVERSY

The reason we emphasize this issue of how an animal is identified is that a dog's worst enemy can be owners who insist on bragging about their "wolf-dog." The state of Michigan defines a "wolf-dog" as any animal so identified by its owner, regardless of its true ancestry. The take-home message from this is that you should be very careful how you describe your animals, especially when dealing with animal control, other law enforcement, or even strangers. All dogs may be wolves; however, owners who identify their dog as a "wolf" or "wolf-dog" could create problems for animals that are clearly just domestic dogs.

This becomes a major source of contention in court cases. Law enforcement and almost all so-called experts look primarily or exclusively at external physiognomy and ignore, or at least downplay, behavior in their decisions. In Pierotti's experience, this means that once he establishes to his satisfaction that the animal in question is not dangerous—that is, it is well socialized—he must look for physical cues that can convince a judge, jury, or prosecutor that the animal is enough of a dog to be considered within the definition that allows the owner to keep his or her companion. Fortunately, behavioral evidence is admissible, and Pierotti need not rely only on physical features.

This is where cultural traditions and considerations become paramount. Authorities need to be shown that the animal whose fate they may be deciding fits their own concept of what a dog is. Opponents of wolf-dog ownership are aware of these problems and try to develop sets of criteria that they claim distinguish wolves from dogs (these were used in the Massachusetts case discussed above). The problem they face is that all of the criteria they use can be seen in one or more breeds of wolflike dog. As we pointed out in chapter 6, dingoes are considered to be a form of domestic dog, even though they live primarily in the wild in Australia, New Guinea, and parts of Southeast Asia.

In a recent case in Florida, it was not even necessary for Pierotti to make an appearance in court; all that was required was to send the prosecutor a set of photographs that showed the doglike characteristics of the animal in question, accompanied by photos of well-recognized types of dog that were more wolflike than the animal in the case, and all charges were dropped. In this case, the primary issue creating the legal problem was the owner had been going to flea markets, charging people money to have their picture taken with a "wolf." As stated above, in its campaign of wolf-dog persecution, the state of Michigan uses the owner's identification of the animal as a wolf or wolf-dog as its primary criterion for prosecution. No matter what kind of dog you have, it is, in its essence, a wolf. You can be proud of your animal's wolflike traits, but don't let your ego endanger a being you may love like one of your own children by exaggerating its identity.

One major issue that results from the confusion about whether domestic dogs and wolves are the same species is the issue of whether vaccines known to be effective in dogs are also effective in wolves. Most physiologists and many knowledgeable veterinarians assume that rabies vaccines are effective across mammals in general, and certainly across all members of the same species, including all canids. As an example, one brand of killed rabies vaccine, IMRAB, is approved for use in dogs, cats, ferrets, cattle, horses, and sheep (Merial 2008), so it is pretty apparent that it works on a wide range of mammals.

The loophole upon which many opponents of wolf-dog crosses, including many veterinarians, depend is that vaccines are usually tested on a species-by-species basis. This is most important when the vaccines are based on immune responses to live microbes (partially killed), where there is some chance of "breakage," which means that the vaccine may not work against and may even stimulate the infection it is designed to prevent. This situation is much less likely with "killed" microbes, such as IMRAB, used to generate an immune response where there is no chance of breakage.

In 1994, Pierotti participated in a USDA advisory panel on the issue of whether the rabies vaccine should be approved for use in all *Canis*. The scientists on the panel, including individuals from the Centers for Disease Control, several university faculty, and experts from the Smithsonian and U.S. Fish and Wildlife Service, were unanimous in recommending approval of use of rabies vaccines in all canids, including wolves and wolf-dogs, and even coyotes. We left the meeting feeling that we had accomplished something scientifically valid, only to have the USDA veterinarian claim that we had not, in fact, recommended approval. His logic was that he asked us about approval of distemper vaccines in addition to the rabies vaccine. We all agreed that evaluating distemper vaccines was not a human health issue and thus was not the purpose of our panel. According to him, this made our recommendation about rabies "inconsistent." This was one of the most egregious manipulations and misuses of scientific evidence we have ever witnessed, and shows that confusion about the identity of dogs and wolves is not confined to arcane debates between archaeologists.

Pierotti testified at a city council meeting considering an "exotic animal" law in which a Humane Society representative argued that dogs had been tested using rabies vaccines but wolves had not. One of the city council members asked the Humane Society representative about ferrets. His response: "Yes, ferrets are approved. They are related to skunks, in which the vaccine has been tested." This statement was misleading as well as being terrible science: ferrets are an Old World mustelid and skunks are New World; the two are not even in the same subfamily, let alone species, whereas dogs and wolves are the same species. This nonsense continues, even among some people who regard themselves as scientists, because of ignorance and the persistence of the outdated idea that all dogs share a single evolutionary line, separate from their wolf ancestors. More to the point, this is an example of how creationist-style thinking creeps into the cultural traditions even of people who consider themselves educated and progressive, and how even humanism derives directly from Christian religious traditions (Gray 2002). Arguing that vaccines tested in dogs may not work in wolves assumes that the ancient Linnaean classification scheme trumps contemporary evolutionary thinking, implying that dogs and wolves were "created separately."

There is little doubt that rabies vaccines work in all members of the genus *Canis*. Wildlife biologists put out bait containing rabies vaccine for coyotes (*Canis latrans*) in areas where they fear rabies outbreaks. In addition, the recovery program for endangered red wolves (*Canis rufus*) vaccinates all animals before releasing them into the wild, as do many investigators studying gray wolves, such as the Yellowstone reintroduction (Schullery and Babbitt 2003).

Despite this, the Humane Society of the United States (HSUS) opted for a Linnaean creationist perspective, opposing legislation and regulations that acknowledge the scientific consensus that wolves and domestic dogs are members of the same species. This argument has been taken seriously at the USDA and state legislative level (see People of Michigan 2000), despite the fact that it has been recognized in the scientific literature since the 1980s that domestic dogs should be classified as *Canis lupus* (Honacki, Kinman, and Koeppl 1982; D. E. Wilson and Reeder 1993). The state of Michigan employs the definition "Dog means an animal of the species *Canis familiaris* or *Canis lupus familiaris*," even though the former name is no longer scientifically valid.

As a corollary to this effort, HSUS and the American Veterinary Medical Association (AVMA) joined forces to oppose the use of rabies vaccines in wolves and even in crosses between dogs and wolves, on the basis of the argument that these are different "species," even though these organisms are genetically and physiologically indistinguishable and the rabies vaccines approved for use in dogs are also considered to be viable for cats, ferrets, and cattle. The AVMA advised that wolf and wolf-dog crosses should not be eligible to be added to dog vaccine labels (American Veterinary Medical Association 2001). Following a short discussion of the USDA action, Dr. Bonnie V. Beaver, a veterinarian specializing in animal behavior and a member of the AVMA Executive Board, stated, "While veterinarians recognize that having an approved rabies vaccine for wolves and wolf hybrids is desirable, this proposal could have had a significant, negative effect on public health by eliminating the USDA's own requirement of proving rabies vaccine efficacy through direct virus challenge." The AVMA contended that making crosses between wolves and dogs, and their descendants, eligible for rabies vaccination would have set a serious legal precedent by allowing wolves and wolf-hybrids to be called dogs. Dr. Beaver stated: "Taxonomy classifies dogs and wolf-hybrids as subspecies of wolves, not the other way around" (cited from a transcript of the AMVA meeting in Pierotti's possession).

Dr. Beaver's statement and the level of ignorance about evolutionary biology it reveals have serious consequences. "Taxonomy classifies dogs and wolf-hybrids as subspecies of wolves, not the other way around" is wrong at several levels. To begin with, dogs are not a subspecies of wolf; both dogs and wolves are members of the species *Canis lupus*. Even if we accepted her premise that dogs are a subspecies of wolf, however, they would still be considered conspecific and thus would qualify for the rabies vaccine. To continue down this path, assuming that wolves and dogs are separate species, wolf-dog hybrids would not be a subspecies of wolf; no scientist cognizant of evolutionary biology would argue that a hybrid or group of hybrids constituted a subspecies. Hybrids are a result of two different species interbreeding, and have no taxonomic status unless they start to interbreed with each other and form a new and distinct lineage (Pierotti and Annett 1993).

As we have pointed out, there are so many different phenotypes of domestic dog that to call them all one species or subspecies distinct from wolves is misleading; the only consistent link that many breeds of dog have with one another is through their wolf ancestry. Crosses between a wolf and any breed of dog are not hybrids because both parents are members of the same species. Dogs are not a species or a subspecies because they are polyphyletic (Frantz et al. 2016), and they have no consistent set of traits that distinguishes them from other related groups of canids; dogs range in size from 1 kilogram to over 100, and they vary in coat color, behavior, tooth size, skull shape, and so on.

Veterinarians are not trained as scientists; they are trained as professionals in animal medicine and rarely engage in research or read the primary scientific literature. This is why their degrees are DVM, for doctor of veterinary medicine. Many people outside of academia often fail to realize that veterinarians, like medical doctors (MDs), are graduates of professional schools that do not train them in research or put an emphasis on keeping current with new developments in biology, particularly fields like evolutionary biology and animal behavior. In contrast, individuals trained in research who make original contributions to scientific knowledge are awarded the title PhD, or doctor of philosophy. In Pierotti's experience as an expert witness, most veterinarians admit that they are not trained as scientists, and that they have no knowledge concerning evolutionary processes, including genetics and systematics, which involve how animal species are identified and classified. The statement by Beaver reveals this ignorance of evolution and systematics, which has replaced taxonomy in modern science.

CHAPTER TEN

Living with Wolves and Dogs

ISSUES AND CONTROVERSIES

BY NOW MANY OF YOU are probably asking, if humans enjoyed such good relationships with wolves, why don't we still see this today? The short answer is that we do, but very rarely. This chapter and the next represent attempts to investigate contemporary versions of such experiences, first showing negative outcomes and, in the next chapter, positive ones. We start by discussing relationships evidencing conflict involving human, dog, and wolf as well as some truly tragic stories (at least for the canids involved) about humans who tried to establish close relationships and failed through their own fearfulness and the sense of learned helplessness that can be generated by contemporary American culture.

One issue worth establishing up front is that all scholars, including ourselves, make a point of emphasizing how the most tame (or at least readily socialized) of wild wolves alive at the time of early human-wolf interactions were probably the ancestors of the domestic canids that we now identify as dogs. What goes undiscussed for the most part, at least by scholars studying the origins of domestic dogs, is that the persistent campaigns of persecution directed against *Canis lupus* by people of European ancestry (McIntyre 1995; Grimaud 2003; Coleman 2004; Rose 2011; Pierotti 2011a) resulted in strong selection favoring the survival of the shyest and least socializeable of wolves alive today in the wild. Oddly, many investigators writing about dogs, including those we greatly respect, such as Morey

(2010) and Shipman (2011, 2015), write as if today's shy, fearful wolves are typical of the ancestral state of all wolves. They proceed from this assumption to explanations of why contemporary wolves are "dangerous," "vicious," "aggressive" (truly misleading characteristics of these shy animals), and other terms that imply at least some tacit approval of the negative images of wolves that dominate much of popular culture. This is especially true in film and television, where wolves are almost universally portrayed as vicious, unstoppable killers, including films as different as the violent Liam Neeson fantasy *The Grey* and Disney's *Beauty and the Beast,* where the titular beast proves his mettle by defending Belle from a pack of ravening wolves.

It has been well established by those who study the philosophy and sociology of science that many practices within science are value laden rather than value neutral, as they would like to claim (Hess 1995; Tauber 2009; Pierotti 2011a, Medin and Bang 2014). In few areas of research do Western societal values impose themselves more strongly than in the study of wolves and dogs and the relationship between these domestic and nondomestic forms of one species. If human interactions with wolves changed significantly over time, however, it is likely that the wolves that were selected for as companions of humans in the early stages of domestication were under different selection pressures and showed different behavioral traits compared to contemporary wolves living with modern society. During most of human history, we sought out and interacted with the most socializeable wolves, hunting with them, sharing food, and even living in close proximity. Somewhere around 1,000 years ago, this changed. The Christian Church decided that many humans were living too close to nonhumans on a respectful basis, including portraying them on coats of arms. Therefore, the church began to associate bears and wolves with the devil to discourage this practice (Pastoureau 2011). Grimaud (2003, 96) describes a devil with a wolf's head in a Roman Catholic cloister in southern France, in stark contrast to pre-Christian imagery that characterizes Apollo as Lukaeios (born of the wolf), and stories of how the founders of the Persian and Roman empires, as well as Genghis Khan, were alleged to have been raised by female wolves. Persecution by Christian clergy generated cultural changes in attitudes toward familiar European predators. This in turn resulted in mass extermination of wolves in Europe, creating strong selection pressure for behaviors that helped surviving individuals avoid contact with humans.

Once Europeans arrived in North America, they conducted similar campaigns in the newly colonized lands (these lands were not "discovered" by Europeans). Pierotti described the relationship between Europeans and wolves in a previous work:

> Wolves, including gray wolves and red wolves, Canis rufus, used to be distributed throughout most of North America, ranging from Florida to Alaska, and from Newfoundland and Labrador to Texas, New Mexico, and Arizona (McIntyre 1995; Coleman 2004). Today the red wolf is officially extinct in the wild, and the gray wolf in the United States survives only in Alaska, Canada and in small relict populations in Michigan, Wisconsin, Minnesota, Montana, and Idaho (Idaho recently passed legislation designed to re-exterminate wolves in that state). Sporadic attempts at reintroduction in Idaho and Wyoming have met with strong resistance from farmers and ranchers who invariably invoke the image of wolves taking children, even though there are no verified records of wolves in nature ever killing a human being in the history of North America (McIntyre 1995; J. Marshall 1995; Coleman 2004; Mech 1998). . . . Instead, the true relationship between Europeans and wolves is best described as a campaign of unimaginable viciousness directed at wolves by Europeans and their descendants (McIntyre 1995; Coleman 2004). Dates of the exterminations of wolves followed the roots of the Enlightenment (Coates 1998). Wolves were exterminated . . . over almost all of the United States by 1950 (Alaska did not become a state until 1959). This extermination effort did not involve simple shooting, as was the case with bison. . . . Treatment of wolves by European immigrants was exceptionally vicious and appeared to be motivated by revenge (though for what is unclear), fear, and hatred (Coleman 2004). In 1814, Audubon describes a Kentucky farmer torturing to death a captured wolf, and farmers and ranchers regularly nailed heads of wolves to public buildings as though they were captured and executed criminals (Coleman 2004). This behavior continues to this day in the treatment of coyotes, Canis latrans. (Pierotti 2011a, 52)

This persecution of wolves and heavy selection pressure on current populations does not change the fact that historically in North America and other parts of the world, humans valued large independent wolves as companions.

The Coppingers (2001), along with other scholars, assume that the first domesticated wolves were smaller, more fearful, and more easily controlled than contemporary wild wolves (Crockford 2006; Morey 2010; Shipman 2011, 2015). If you are looking for a powerful, intelligent hunting partner, or better yet a pack of them, you don't want the smaller, high-strung sort of animal that contemporary Americans prefer as domestic dogs; you want what you have in the actual wolf—a powerful ally or companion, not a subordinate.

How did humans decide which wolves they wanted to associate with? Or did the wolves perhaps decide? Wolf behavior expert Benson Ginsburg states: "It is my experience that if you put your hand into a pen with newborn wolves, a certain percentage will come immediately and never want you to leave. . . . Other pups will run away and still others will be stuck in avoidance-approach. As adults, the social ones can become sociable to humans" (quoted in Derr 2011, 86). This statement is very accurate, as individual cases in chapter 11 exemplify. There is an account from an eighteenth-century European explorer about how Indigenous Americans would visit a wolf den and play with the pups, while the adult wolves watched, and mark some of these puppies with ochre (Hearne 1958, 240), which Hearne interpreted as an odd sort of ceremony. Another explanation, however, is that the Indigenous people wanted to be able to recognize the most social individual pups, the ones most likely to establish relationships with humans, when they matured. Ochre lasts a good while when rubbed into skin and hair, and its use might have functioned as Indigenous equivalents of tagging.

Over his lifetime, Pierotti has dealt with a couple of dozen wolves, around a hundred crosses between wolves and dogs, and a couple of hundred dogs in various capacities, including as owner, companion, scholar, mentor, trainer, educator, and expert witness. Pierotti has been charged more than a dozen times and bitten a half dozen; in every instance the animal interacting with him was a large, male, unequivocally domestic dog, usually from an AKC-recognized breed, including German shepherds, Rottweilers, and Belgian sheepdogs. He has never been charged or bitten by a nondomestic wolf.

The social bond between humans and wolves that have been changed into domestic dogs is a source of both major pleasures and major conflicts between humans and their nonhuman companions. Large domestic dogs have the anatomy of serious predators combined with a confidence in their interactions with humans that can lead to aggression. In contrast, wolves and high-percentage crosses between wolves and dogs tend to be more timid and retreat when interacting with unfamiliar humans for the reasons described above. One point raised by Dr. Noreen Overeem, a veterinarian with more than thirty years of practice, is that Ginsburg's observation concerning variation in personalities within litters of wolf puppies can be applied to dog litters as well—some puppies are friendly and confident, others are aloof, and some are very shy and fearful. According to Dr. Overeem, who works with dogs, wolves, and crosses as part of her practice, dogs in these latter two categories present many of the behavioral problems and even crises that occur within human/dog relations.

This point is borne out by data; shy, aggressive, and fearful members of recognized dog breeds are the ones that end up attacking and even killing humans, especially children. The U.S. Centers for Disease Control (CDC) reports that more than 300 people were killed by 403 dogs in the period from 1979 to 1996; of these, only fifteen incidents involved purported wolf/dog crosses, and some of these are highly questionable (see Nickerson case below). The most dangerous "breed" by far (118 deaths) was "pit bull types," both purebred and cross-bred; second were Rottweilers with 67, third was mixed breed, and fourth was German shepherds. One study, which purports to represent "research into the most dangerous dogs . . . performed by the American Veterinary Medical Association (AVMA), the CDC, and the Humane Society of the United States (HSUS)," provides a list of the ten most dangerous breeds: (1) American pit bull terrier, (2) Rottweiler, (3) German shepherd (Alsatian), (4) Siberian husky, (5) Alaskan malamute, (6) Doberman pinscher, (7) chow, (8) presa Canario, (9) boxer, and (10) Dalmatian (Sacks et al. 2000). This group includes three spitz-type dogs (malamute, husky, chow), and the German shepherd, all well-known breeds recognized by the American Kennel Club. Five other AKC-recognized breeds are mostly Molossers, large dogs with short snouts and powerful jaws (pit bull, Rottweiler, presa Canario, boxer), which are responsible for the vast majority of fatal attacks (http://www.dogsbite.org/dog-bite-statistics-fatalities.php).

The general lesson is that large dogs present the greatest risks, especially those that combine strength with neotenic personalities; they are protective of social bonds with humans, but social immaturity renders them aggressive toward unfamiliar humans and dogs. It is important to note that if we consider simply dog bites rather than fatalities, many popular smaller breeds are implicated, although attacks by small dogs rarely result in fatalities. This is especially true with smaller children, whom domestic dogs feel they can dominate or drive off. Children do not respond like subordinate dogs or puppies would in such situations— rolling on their backs, exposing their bellies or throats, and urinating on themselves—and as a result of this behavioral incompatibility they may be bitten, injured, and on occasion killed.

Pierotti appeared in several cases where it was alleged that an animal purported to be a wolf-dog supposedly attacked a child. If you noted the number of qualifying words in the preceding sentence, you will appreciate how public attitudes and laws get twisted around. In such cases, two things need to be established: first, whether the animal involved actually has any recent wolf ancestry (more so than any easily recognized domestic dog, since all dogs are wolves), and second, whether an attack actually took place. A third question that has arisen in some cases is whether the animal characterized as a wolf or wolf-dog was even really involved in the alleged attack.

We start by discussing a case that involved all three variables. In 1994 in western Missouri, a five-year-old boy carrying a plastic knife noticed a piece of board missing in a six-foot-high wooden fence. When he peeked through the gap, he saw dogs (note the plural) inside the enclosure. Being five years old and male, he decided it might be interesting to stick his hand holding the knife through the opening to challenge the dogs. When he did this he was bitten, or at least his hand was grabbed by a canid mouth. As he pulled his hand back, he sustained a couple of deep cuts. Whether this would have happened if he had remained calm is open to question, but we are discussing a five-year-old boy.

The child was taken to the hospital; damage to his hand required minor surgery and numerous stitches. His parents (who had the child while they were still in high school) called the police. Police investigators found that the person who owned the dogs described one of them as a "wolf hybrid," possession of which is a violation of Missouri law. At this point the state of Missouri's Department of Conservation, which is in charge of "wildlife," called Pierotti in as an expert witness to confirm the animal's identity as a wolf hybrid. When the conservation officer took Pierotti to the cage where they were holding the animal, he was startled to find a very obvious domestic dog, trembling and fearful, not at all aggressive. This animal was a heavy-legged, overweight dog that looked like a cross between a German shepherd and a yellow Labrador retriever. Pierotti's initial response to the conservation officer was "You've got to be kidding me." Topping this off, the conservation officer admitted that authorities were not even sure which dog had "bitten" the child because there were three in the yard and the boy had not actually seen his "attacker." Therefore, not only was the accused animal not a wolf hybrid of any kind, but law enforcement was unsure if they had the right animal. They were willing to prosecute, however, because the owner had identified his pet as a "wolf hybrid," and the authorities involved were so poorly prepared in identification skills they could not accept the evidence of their own eyes.

This case attracted attention in the local news media because of the "wolf" angle. The father appeared on television, ranting about how his child had been "attacked by a vicious and aggressive wild animal" and, as always on local newscasts, "If it bleeds it leads." One reporter called Pierotti at his university office and asked his opinion. Pierotti told him that the animal was in no sense a wolf, no more than any other domestic dog drawn at random, which the reporter reported on the local news. Pierotti was quickly contacted by the conservation officer, who informed him that because of his "obvious bias" the state would not call him as a witness after all.

Pierotti was subsequently contacted by the defense attorney, who did want his testimony as an expert witness. Technically, as part of the state's case, Pierotti had to be subpoenaed to ensure he got the chance to present his opinion in court. In court he was able to clarify that the animal was not a "wolf hybrid" of any kind, and that the animal was timid and fearful rather than vicious and aggressive. The charges of owning a wolf hybrid were dismissed, but the owner was convicted of misdemeanor animal neglect, supposedly because the gap in the fence indicated a "lack of control," defined as "to reasonably restrain or govern an animal so that the animal does not injure itself, any person, any other animal, or property" (*State v. Choate* 1998).

This weak, compromised conviction was overturned by the Missouri Court of Appeals on the grounds that "there was no evidence . . . that the dog could have inflicted this injury on the child without the child inserting his arm into the enclosed area. The opening in the fence was not shown to be of sufficient size to allow the animal to injure any child absent the child intentionally exposing himself to the danger" (*State v. Choate* 1998).

In 1996, Pierotti was called in as an expert witness in a similar case in North Carolina. Another teenage father with an eighteen-month-old boy had been visiting a facility where some wolf-dog crosses were kept. In this situation there was little doubt concerning the identity of the animals, and Pierotti was able to confirm this. What transpired, therefore, was a civil case concerning "viciousness" and "aggression." The nineteen-year-old father had dangled his young son in front of an agitated animal, even holding him against the ten-foot-high chain link fence. The animal's snout was too big to fit through the fence—but the child's foot was not. What happened could easily have been predicted: an animal grabbed the child's foot. The father began pulling on his child, and in the ensuing tug-of-war, the child lost his little toe. The police investigated, but under the circumstances the officers declined to press any charges (although child endangerment by the father was an option).

The parents resorted to a civil suit, even though the wolf-dog's owner had offered to pay all medical expenses. Pierotti was brought in to evaluate the animal and then to give testimony for the defense. Pierotti visited the owner and his animals, and as usual insisted on being allowed to enter the enclosure. There he was able to observe that the animal involved in the incident was shy and tended to retreat from unfamiliar humans. It was not in any sense aggressive or vicious but, as a normal modest-percentage wolf-dog cross, it was likely to snap defensively out of fearfulness if cornered.

When Pierotti was called to the witness stand, the plaintiff's attorney questioned his qualifications, on the grounds that his university position was "evolutionary ecologist" rather than "dog behaviorist," a position the attorney seemed to assume most universities keep on staff. The reason this point was raised is that the attorney, and his clients, paid around $3,000 (plus expenses) to an animal trainer, certified as a "dog behavior expert," who spent five minutes observing the animal from outside the enclosure and declared it "dangerously aggressive." Pierotti's publications in the field of animal behavior convinced the judge that he should overrule the plaintiff's objections and that he could in fact testify. He presented his testimony, and the first question he was asked during cross-examination was "How much are you being paid for this testimony?" Pierotti gave his usual answer: "I only accept expenses. I charge nothing for my work; I am employed as a university professor. Testimony like this is considered service to the greater community." After this testimony, the defense rested. The jury retired, coming back with a finding for the defense within three hours. All the plaintiffs received were medical expenses—exactly what they had already been offered before they initiated their lawsuit.

A third case of this nature, concerning putative attacks on small children, took place in a Detroit suburb, where a "wolf-dog" was accused of attacking and biting a three-year-old girl. This situation was complicated by the fact that the owners of the animal were selling its offspring as "high-percentage wolf puppies." The circumstances that led to the incident involved the latest litter of these puppies. It was a mild February day, and the daughter of the animal's owners had a friend over to play in the yard. Neither parent was home; a grandfather, who was watching the girls, released the three-week-old puppies to play with them. It was the first time these puppies had been in the yard with a person outside of the family, but girls and puppies romped together. For reasons best known to himself, the grandfather then decided to release the puppies' parents. At this point something happened that resulted in the visiting girl getting "bitten" by the puppies' father. We place this word in quotes because of findings that took place after Pierotti became involved in the case. At the time, however, the girl's mother was called, the police became involved, the animal's alleged wolf-dog identity became known, and the dog was confiscated by the municipality. The owners of the animal were charged with possession of a dangerous animal, and if found guilty, they would be fined and the animal would be put down.

The prosecutor in this case was a young woman who was determined to win. She had the animal's head X-rayed and sent the X-rays to Danny Walker, state of Wyoming assistant state archaeologist and adjunct professor of anthropology at the University of Wyoming for analysis and his opinion of the animal's identity. She was inexperienced and had no idea how such analyses need to be conducted; therefore, she neglected to have a reference scale X-rayed next to the skull or to provide any measurements of the animal, which provided misleading information concerning its size. She also "researched" the rabies vaccination situation, operating under the Michigan premise that wolves are a different species than dogs (see People of Michigan 2000); hence, she argued based on the erroneous assumption that the rabies vaccine could not work in wolves.

Pierotti flew to Detroit the day before the trial was scheduled and was able to examine the evidence in the case, including photographs of the "victim" of the "attack." These revealed that the "bite" in question consisted of a single small cut in the middle of the girl's back. Since every carnivore in the world typically has two upper and two lower canine teeth, it is very difficult if not impossible to bite and leave only a single break of the skin. Moreover, it is almost impossible to bite the middle of the human back (try it). Pierotti was also able to see the X-rays and read Dr. Walker's report. He was surprised to find that Walker had actually attempted an analysis despite the lack of an appropriate scale for size. The X-rays revealed that this animal had the usual two upper and two lower canines, which rendered the "bite" diagnosis questionable.

The next morning, before court convened, Pierotti visited the facility where the animal was quarantined and examined it for an extended period. As with the Missouri case, Pierotti's initial reaction was that he was being kidded. This was a small animal, no more than fifty pounds, with medium-length silky fur, a prominent stop (forehead), and showed no more resemblance to a wolf than any other dog of that size with upright ears and an uncurled tail. The animal was initially shy (it was being held by unfamiliar people), but he allowed Pierotti to pet him and conduct a physical examination, and thus Pierotti found what he suspected—the dog had untrimmed sharp dewclaws.

The dewclaw, basically a canid's thumb, is often employed when a canid uses its paws to attempt to control something. Pierotti has sustained many more dewclaw cuts over the years than bites, because canids typically use their paws during rambunctious play. This led him to conclude that the "bite" the girl sustained was, in fact, a dewclaw cut, which cast an entirely different light on the dynamics of what actually happened. Even though it was February in Michigan, everyone agreed that it had been a mild day: so mild, in fact, that all the girl had been wearing over her torso was a loose-weave sweater over bare skin, which would have readily allowed the dewclaw access to her skin. Pierotti's interpretation was that after the male dog was released, he found an unfamiliar child with his young puppies. He rushed over, knocked the girl down, and put his paw on her back to control her movement. Being a human child rather than a well-mannered puppy who understood social rank, she squirmed and struggled. This led the male to place more weight on her to control her movement, resulting in the dewclaw breaking her skin as he pressed down with his paw. This was not an "attack" but a male parent trying to control a possible threat to his offspring. If the girl had lain quietly nothing more would have happened to her other than perhaps her clothes might have gotten dirty. Instead the usual hysteria ensued, a "vicious" attack was alleged, and an animal's life was at risk because he was protective of his puppies.

The prosecutor was unhappy with Pierotti's interpretation of the animal's identity and with his interpretation of what actually occurred. She asked Pierotti to explain why he disagreed with Danny Walker's identification of the animal as a "possible wolf-dog." Pierotti pointed out that her failure to provide a scale led Walker to presume that the animal was the size of typical wolf, but that its actual small size, combined with its silky fur, precluded its being a nondomestic form of wolf. She asked Pierotti the typical question about how much he was being paid and Pierotti gave his usual response (it turned out she had paid Walker more than $1,000 for his expertise). She then asked about the rabies issue, to which Pierotti replied that this animal was clearly a domestic "dog" and had been vaccinated—but since no actual bite was involved, the point was moot anyway. When Pierotti pointed out that killed viruses were used for a wide range of mammals, out of frustration she asked if Pierotti was paid as a representative by IMRAB, which was clearly not the case, and the judge admonished her. As she wound down, the judge turned to Pierotti and said he had one question: "Dr. Pierotti, do you consider this to have been a predatory attack?" Pierotti replied, "Your honor, if this had been a predatory attack, none of us would be in doubt about the outcome." Hearing this, the judge nodded his head and suggested that the defense attorney take Pierotti to lunch.

The defense attorney said that this final exchange had "won the case," which turned out to be an accurate assessment. That afternoon the judge dismissed all charges and released the animal back to his owners, who were happy until Pierotti told them that although he had not asked for a fee, he did have a strong request: that they cease advertising the puppies they sold as "wolf-dogs." Pierotti pointed out that this type of careless action had nearly cost them an animal they cared for deeply. After extensive discussion, they promised that they would accede to this request.

These cases all had relatively positive outcomes, but it is essential to emphasize that such situations often have serious negative results, as happened in a case in which Pierotti was only peripherally involved after the fact. On March 2, 1989, in the small Michigan Upper Peninsula village of National Mine, five-year-old Angela Nickerson got off of her preschool bus at 11:30 and found no responsible adult at home (her aunt was in bed with her boyfriend, but they were so distracted that they "did not hear anything"). Playing outside, Angie drew the attention of at least two dogs, subsequently identified by local police as a male eleven-month-old 110-pound "husky" and a female German shepherd/husky mix. The former had recently been given to Angie's aunt by her boyfriend, who adopted it from an animal shelter. Sometime between being dropped off by the bus at 11:30 and 12:00, Angie Nickerson was attacked and killed by these two dogs. (The police report on Angie's death is available at https://goo.gl/8oFLfn. Be warned: this report is graphic.)

Several years after this tragic incident occurred, Beth Duman, BA, a self-described "Wolf Lady" and "wolf expert," former high school science teacher, dog trainer, and anti-wolf and anti-wolf-dog advocate (http://www.casinstitute.com/beth.html), heard about the Nickerson case, and apparently decided it presented an opportunity to generate publicity for her cause, despite the fact that everyone involved with this case had identified Ivan, the animal that seemed to be the primary attacker, as a dog—a husky or malamute. Duman herself never saw Ivan, alive or dead; he was cremated after being shot by the police and she depended on photographs of his body for her identification.

Beth Duman has an interesting history in regard to her relationships with wolves and wolf-dogs. We first became aware of Duman in 1994 from a *Smithsonian* magazine article titled "Wolves and Wolf Hybrids as Pets Are Big Business—but a Bad Idea" (Hope 1994), one of the first critical attacks on the practice of having wolves as companions. Duman was featured prominently in this article because she and her family had a supposedly pure wolf as a pet for four years. As Duman describes the situation, "She was the sweetest creature with kids, with my husband, and me. . . . As a biologist and champion of wolves I used to take her around to schools to teach kids just how lovely wolves really are." Around the animal's fifth birthday, the following occurred: "One spring afternoon Bob [Duman's husband] and I were in the backyard pen with her, petting and scratching her when—wham!—without any warning she was up on her hind legs, her front paws on Bob's shoulders, pushing him back against the fence, sinking her canines deep into his right forearm. It was all over after that, Bob could no longer go in the pen because she would attempt to attack him every time. We had to get rid of her." According to Duman, her husband had hurt his lower back and "that's all a wolf has to see, some little chink in your armor—and it'll be in your face like it would be with another wolf" (39). This is an odd statement from a self-described champion of wolves and an expert on wolf behavior because, as described by Mech (1999), wolves rarely attack one another, especially within a social group. In Pierotti's experience such events do not occur unless the animal has been mistreated at some point by the person it "attacks," but this point was not even considered in Duman's account or in the article.

Duman provided a more detailed and nuanced account of the interaction between Bob and the animal in her presentation before the AVMA (Duman 1994), which we reproduce here exactly as Duman personally transcribed and circulated it from the transcript "Usually it seems that the challenges need to build up by little steps and even when why husband was bit, bitten, and when we looked back in time, there had been some minor things that had happened, that probably led up to the big attack. But we didn't even realize that they had happened. Like he had been chased out of the pen once over a husky puppy that was living with here. Things likes, so UUUUSSSSSUALLY YOU have a little time to do things before they get out of hand."

In many ways this recalls Ray Coppinger's account of his interactions with Cassi at Wolf Park, described in detail in chapter 1. An "expert on wolf behavior" should not have failed to recognize the potential importance of "minor things that had happened." In our experience, you do not tolerate "minor things" such as being "chased out of the pen," and it is never made clear exactly how the "husky puppy" fits into the dynamic.

In any event, after this experience, Beth Duman appears to have become convinced that if she could not trust a wolf as a pet then no one could. She received "training" at Erich Klinghammer's Wolf Park, declared herself an expert on "wolf" and "wolf-dog" identification and, with an endorsement from Klinghammer (she identifies herself as Wolf Park's "representative in Michigan" [Duman 1994], although exactly what this means is not clear), she set off on a campaign to ban wolves and wolf-dogs as companion animals, no matter how well individual animals were functioning within their human groups.

Duman approached Patti Nickerson, Angela's mother, years after the fact, and convinced her that Ivan was not a "sled dog" but a "wolf-dog," based in large part on his size. Interestingly, in her 1994 presentation before the AVMA, Duman performed a quiz called "Guess the Weight of the Wolf": she put up an image of an animal and asked, "Who's the smartest vet in the room? And no fair if you're real smart. Okay somebody raise your hand and tell me how much that wolf weighs. Come on, wake up. 100 pounds, okay, 110. . . . That wolf weighed 70 pounds" (Duman 1994). The reason we present this detail is that Duman, discovering from the police report that Ivan weighed 110 pounds at eleven months, used this criterion to declare him a "wolf-dog," which is directly counter to the example she presented at the AVMA meeting in 1994. If Ivan had been a wolf-dog, he should have weighed no more than 70 pounds, as Duman herself indicated in her presentation. This sort of inconsistency became a hallmark of interactions with Beth Duman.

Nickerson, apparently guilt stricken and eager to embrace a story line that would take the focus off allegations that she was an unfit parent, seized onto Duman's argument that Ivan was "part wolf." Her daughter's death was not a result of her negligent care, she maintained, but the fault of other members of her family who "ignored her warnings" and of the Humane Society animal shelter that adopted out a "dangerous wild animal" (Nickerson 1997). Nickerson initiated a series of lawsuits, including one against her own parents for "endangering her daughter." Ironically, Eileen Liska, independent lobbyist for the Michigan Humane Society and apparently at one time president and chief executive officer of the Humane Society of the United States, was trying to get an anti-wolf-dog law passed. Liska involved Beth Duman and Patti Nickerson in this campaign, despite the fact that at the time Nickerson was suing one of the Humane Society's animal shelters.

Pierotti became involved in the issue when he testified at a state Senate committee hearing in East Lansing in June 1997. He explained that there were a number of AKC-registered breeds that could be easily mistaken for "wolf-dogs," and also that the language of the proposed legislation regulating ownership of *Canis lupus* would effectively ban ownership of all dogs, given the reclassification of all dogs as *Canis lupus*. Nickerson's testimony followed. She screamed and cried while showing graphic photos of her daughter's autopsy to the panel.

After the hearing, Nickerson cornered Pierotti in the hallway and accused him of being "one of the ones responsible for her daughter's death." He replied that he was sorry for her loss, but that there was no evidence of any kind that her daughter had been the victim of an attack by a wolf-dog. Pierotti pointed out that according to Beth Duman's own logic and his own extensive observations, no wolf-dog could weigh 110 pounds at the age of eleven months because of their slower rates of growth and development relative to domestic dogs (Morey 1994). Pierotti emphasized that a wolf or wolf-dog of that age would weigh about 60–70 pounds, and it was clear from the police report that this was not the case with Ivan. Nickerson had testified that she had regarded Ivan as dangerous (Nickerson 1997).

Pierotti's testimony was helpful in stopping passage of the bill in this early incarnation, and he corresponded with a couple of state senators who thought the bill was unnecessary because he had convinced them that "Dangerous Dog" laws already in existence covered the situation without creating breed-specific legislation, which invariably leads to more confusion. Eventually a version of the bill—now called "Angie's Law" in memory of Nickerson's daughter—pushed by Duman, Nickerson, and the Michigan Humane Society, and sponsored by Glenn Shugart (R-Kalamazoo), Glenn Steil (R-Grand Rapids), Joel Gougeon (R-Ogemaw), and Mike Goschka (R-Larkin Township), was passed in 1999, going into effect in July 2000 (State of Michigan 2000). This illustrates how conservative politicians ally with supposed progressives to push anti-wolf and anti-wolf-dog legislation. If you read this bill, you will find that the only way a "wolf-dog" can be identified is by consulting an "expert on wolf-dog identification" such as Beth Duman.

The next animal Pierotti was asked to evaluate in Michigan was a tall, long-legged, long-snouted, pale-colored, ninety-pound German shepherd/Alaskan malamute cross about two years old (dogs of this cross superficially resemble wolves more than either parental strain; see figure 9.2). The owner's ex-husband had told a neighbor that the animal was a wolf-dog, which was then reported to local animal control and the identity was "confirmed" by Beth Duman after the wolf-dog law was passed. The animal was taken from his owner and kept in an animal shelter while his fate was decided.

Despite his superficial resemblance to a wolf, the animal had dark eyes, pink color on his nose, a pronounced stop (forehead), and no dark pelage under his eyes, none of which would be typical of wolf-dogs. In addition, when the animal stood upright, his hind legs extended backward behind his pelvis, a trait found in German shepherds but completely uncharacteristic of wolves. The final evidence was that when Pierotti encountered the animal in its enclosure, he pushed itself up against the wire and solicited attention. Again, this is very different than the behavior of wolves and wolf-dogs when confronted with an unfamiliar human, especially a male. Pierotti was allowed to enter the enclosure after signing a waiver, whereupon the animal continued to solicit physical contact, rubbing and curling up against him. After Pierotti spoke to the authorities, the animal was released back into the care of his owner, where he lived to be fourteen without ever causing an incident of any kind.

The next case in which Pierotti was asked to examine an animal identified by Beth Duman as a "wolf-hybrid," the criteria for identification were much more misleading. The animal in question was a large, 130-pound dog with short heavy legs, relatively short and sleek fur (unlike the thick, multilayered fur of wolves), and a barrel-like body—his only wolflike characteristic was upright ears. He did not remotely resemble a wolf and was probably a mix of black Lab and German shepherd. Beth Duman had described this "wolf-dog" as being thirty-six inches at the shoulder. Pierotti remeasured the dog in the animal control facility and found he was only thirty inches. This may not seem like a major difference, but it means that Duman's measurement was off by 20 percent, and biased in the direction that favored her interpretation. The case was thrown out of court and the animal was returned to his owner.

Incidentally, Beth Duman has never appeared in court to testify in a case in which Pierotti was involved, even though she lists as first among her accomplishments "Expert Witness; Various wolf/dog hybrid court cases in the State of Michigan" (http://www.casinstitute.com/beth.html).

Overall, Pierotti's impression is that the U.S. court system functions reasonably well in such cases, given a judge, or jury, willing to listen to evidence. State legislators can be insightful, but in contrast to judges, too many are easily manipulated by emotional appeals such as those perpetrated by Patti Nickerson. They routinely seize chances to pass laws that may have popular support even if they completely ignore scientific evidence, as in the Michigan wolf-dog law, which is utterly unscientific in its language. This is particularly true of many who argue that dogs and wolves are separate species, adhering to the Linnaean argument that they are the results of separate creation events, which fits in with creationist, anti-evolution beliefs.

ROMANTICISM AND THE CONCEPT OF "WILD"

Anti-wolf influences show up in a number of unusual circumstances. It has been argued that Americans, whether for or against development or environmentalism, tend to operate within the romantic philosophical tradition (Berlin 1999; Pierotti and Wildcat 2000; Pierotti 2011a). Europeans (and Euro-Americans) are easily convinced that wolves are vicious, destructive killers, especially of human children, despite the fact that *no wild—or even purebred captive—wolf has ever been implicated in the death of a child in North America*. We have shown how such language creeps into accounts by scholars involved in studies of domestication. Domestic dogs, especially the Molosser breeds, kill orders of magnitude more children than wolf-dogs have ever been accused of doing (Sacks et al. 2000).

Euro-Americans who consider themselves "pro-wolf" often romanticize wolves. When faced with actual flesh-and-blood animals that want to live their lives on their own terms, however, they become fearful and helpless. Ceiridwen Terrill's book *Part Wild* (2011) is a case in point. Subtitled *One Woman's Journey with a Creature Caught between the Worlds of Wolves and Dogs,* the book is solidly within the romantic tradition. The unaddressed issue, however, is whether Terrill actually had a wolf-dog in the first place. Beth Duman helped convince her that her problems with her animal, Inyo, were not her fault but a result of the dangerous animal she had been harboring. A more objective observer, however, might come away with the impression that Terrill had ineptly raised a young animal from puppyhood, and that Inyo was simply looking for a competent human to care for her and shape her behavior.

Terrill's experience begins after she flees an abusive boyfriend, who apparently killed the dogs she left behind. Looking for a new dog in an animal shelter, she found Cochise, whose cage reads, "male wolfdog hybrid." She feels drawn to the animal's mixture of fear and aggression, which she considers to represent its wildness: "As I watched Cochise I felt an extra tenderness toward him and something broken inside me righted itself" (4). She offers to adopt the dog, whose primary reaction has been to growl and retreat from her. When she is told she cannot, that the shelter is going to put him down, she identifies further with this "abused and abandoned" creature. "I'd read about La Loba, The Wolf Woman, a strong, healthy woman with a fierce independent spirit. I wanted to be fierce like that, to stand up for myself and protect the members of my pack. . . . I was determined to rescue another wolfdog, to be the pack for the one who had none" (7).

Terrill contacts wolf sanctuaries but is unsuccessful in her quest. A friend offers to take her to "see some wolves" that might be breeding soon. She is disappointed, writing, "I'd wanted to rescue an adult wolfdog, not buy a puppy" (2011, 11), which reveals the depths of her naïveté, because it is much easier to raise your own animal from puppyhood rather than take on an adult and its unknown history, which could include serious traumatic experiences. She is taken to a house in the suburbs and shown a three-year-old male she is told is "pure wolf" and a two-year-old female that is allegedly a mix of wolf and Siberian husky, with a tail that curls over her back, suggesting that the husky is predominant, if not the entire pedigree. This pair will be bred to produce the puppies. It is a crucial element of this story that Terrill does not even attempt to check the breeder's veracity, nor does she consult anyone who could advise her about what she is getting into. This is especially important, because a two-year-old dog that might have significant recent wild wolf ancestry is not ready to breed at two years of age, whereas a domesticated dog would be ready. The breeder tells her, "You have to be tough with them. Show them you're alpha" (13). She then ponders the question: "*Why if wolves and dogs are the same species, did they behave so differently?*" (15), which reveals that she knows nothing about the behavior, or even the physical traits, of spitz-type dogs.

Terrill's emotional instability is hinted at repeatedly in her writing. When people ask her why she wants a wolf-dog, she says she loves "that you can't control them." She asserts, "Only a wolfdog had the strength and endurance to keep up with me on wilderness backpacking trips." She adds, "There was more to it than that. Their wildness made them aloof and wary . . . unlike dogs, and unlike me, they couldn't be charmed by some sweet-talking stranger" (2011, 17). This is a person who wants an animal that is fearful and cannot be controlled, thinking this will make it her nonhuman soul mate. She does not regard these animals as powerful independent spirits who want to have rules that make sense to them and consistent structure in their social relationships. Also, if she thinks that there are not breeds of dog that could easily keep up with her on wilderness hikes, this reveals deep ignorance about the endurance of dogs.

While seeking her wolf-dog, Terrill falls for another damaged soul, a severely depressed unemployed man who is more than $100,000 in debt, a fact he neglects to reveal. She adopts a puppy from the breeder and names her Inyo, after the county in California where both the lowest elevation (Death Valley) and the highest (Mt. Whitney) are located. "I knew I would put all of myself in this tiny creature" (2011, 27). The most telling indicator of difficulties to come is described later: "One afternoon when I came out of [a grocery store] . . . with a fresh package of marrow bones, Inyo leaped from the rear seat and tore the package from my hand. . . . When I reached over to snatch it back and redo the whole here's-your-bone-that's-a-good-girl routine, she growled and her hackles pricked. The skin of her muzzle and her canines, sharp as scythes, made me hesitate, exactly as they were meant to" (83). Terrill backs down and lets Inyo keep the bones. Any person with a wolf-dog—or any large dog, for that matter—should have recognized this as a critical point in the relationship, determining who controls the dynamic. She needed to enforce strong discipline at this juncture. As an example, Tracy McCarty and Peter (see Chapter 10) sometimes had power struggles, often over French fries, which were a favorite of Peter's. In every case McCarty did not back down and made sure that Peter recognized that she was in control. In the current case, Terrill's reliance on "positive reinforcement" clearly was not working, and she needed to establish herself as the calm, dominant individual in the relationship.

What Terrill does instead is consult so-called experts, one of whom is "wolf-behavior specialist and educator" Beth Duman, who advises: "That alpha stuff is garbage. . . . This treatment terrorizes dogs and in some cases makes them aggressive. . . . People who work with captive wolves found out a long time ago that if you pick on wolves they'll pick on you back . . . so you'd better let go of that dominance model unless you want to die young" (2011, 84; this is standard Erich Klinghammer dogma—see chapter 1). In Chapter 11 we provide numerous examples that show how wrong this advice of Duman's really is, which also leads to the tragic outcome of Inyo's story.

Terrill follows the account of her capitulation to her puppy's threats with the information that her husband has overdrawn their checking account and drained their savings account. She is then institutionalized for threatening to kill herself when she finds out about her husband's credit problems. She describes her mental state: "I developed nervous habits. Scanning ads online, I would tug strands of my hair until they fell out and scratch my scalp until it bled. My front teeth were loose from months of grinding" (2011, 151). She is so stressed that she forgets to set the brake on her car in a gas station, with Inyo in the back seat. The car rolls into a field 200 yards away, thus providing Inyo with another reason for insecurity.

All of this personal narrative is mixed in with tales about how she traveled the country seeking answers about what she might have done wrong. Interestingly, Terrill seems to seek only people who convince her that wolf-dogs are bad and problematic. As far as we can determine, Terrill never attempted to contact anyone who could give her an objective evaluation. This is unfortunate, considering that one of the best domestic wolf-dog breeders lived within an hour of her in Reno.

At one point, Terrill borrows money from her stepfather to travel to Siberia to interview Lyudmila Trut, the Russian geneticist and physiologist who conducted much of the fox domestication study discussed in chapter 9. Trips to Siberia cost thousands of dollars. If she had spent this amount on constructing an adequate pen and shelter for Inyo she might have had fewer problems, but instead she finds another person who does not understand her. She tells Trut that "American *taxonomists*" had reclassified dogs as a form of wolf and reports Trut's look of disapproval. This simply reveals more of her ignorance. If she had known how to ask the right questions, she could have asked Trut if she considered her "tamed" foxes to be a different species than the foxes used as controls, which would be the equivalent situation to the dog/wolf dynamic. Instead Terrill improvises off Trut's reaction, contending, "Dogs are not wolves. Wolves are not dogs. Evolution and domestication have seen to it" (2011, 231), revealing ignorance of both evolution and systematics.

At this point, rather than provide more depressing information concerning Terrill's behavior, let's examine Inyo and her social situation more closely. We are not describing the fantasy Inyo, who would save Terrill from herself and release her inner La Loba, but impressions based upon the images of Inyo available online and in the few photos in *Part Wild*. Inyo is an interesting animal; however, based on the photos in the book, she does not look all that wolflike. She lacks the spectacles around the eyes that are almost always found in high-percentage wolf-dogs, and she has dark irises, not the gold normally associated with either wolves or huskies, her alleged parental lines. She also has an unusually long tail for a "wolf-dog." It would have been good to examine her in the flesh, but this is no longer an option.

More telling are Terrill's accounts of Inyo's behavior. Aside from Terrill's own reports, there is no account of Inyo snarling at, nipping, or showing aggression toward any human. Nor does Inyo seem to be shy, or retreat when encountering unfamiliar humans. This indicates quite strongly that she is not nearly as "wild" as Terrill would have us believe. One particularly pertinent section can be seen during the period when Terrill is institutionalized. For several weeks while Terrill is under care and away from Inyo, Inyo is cared for by her husband, with no reported difficult situations arising. Given her descriptions of Inyo's actions from the rest of the book, this seems very unlikely if Inyo is in fact the high-percentage wolf-dog Terrill claims.

In Pierotti's experience, the animal Inyo most resembles would be the malamute-shepherd cross in Michigan described above that was identified as a wolf-dog by Beth Duman. If you compare the photograph of Inyo on the cover of the Simon and Schuster edition of *Part Wild* (http://books. simonandschuster.com/Part-Wild/Ceiridwen-Terrill/9781451634822) with images of the malamute/shepherd cross, the resemblance is astonishing (see figures 9.2 and 10.1).

There is no evidence presented that Terrill knows anything about the behavior and social dynamics involved in rearing malamutes, German shepherds, or even huskies (which is how she identifies Inyo when asked). This suggests that she would not have been expected to question the breeder from whom she purchased Inyo about the accuracy of the alleged pedigree.

FIGURE 10.1 German shepherd/Alaskan malamute cross. (Courtesy SSM)

The real issue is, Terrill made up her own romantic story and then seems frustrated when it does not play out according to her fantasies. Inyo was supposed to save her and be her protector, her wild spirit; she forgot, however, that Inyo was actually a young animal who needed guidance, not romanticizing. Once Inyo showed her own spirit, this led to conflicts, which seem endless according to the account provided in the book. As a few examples, first the breeder had suggested that Terrill take two puppies, so that Inyo would not be alone while Terrill was absent or distracted. This was probably useful advice, because one problem Terrill had with Inyo was that she howled when left alone, a perfectly reasonable response for an immature young animal left alone by her caregiver. Inyo also was destructive to furniture and other household objects when left alone, a clear case of abandonment issues that can be seen in many breeds of domestic dog, although Terrill and her advisors treat this as evidence of wildness. Terrill rejected the idea of keeping two animals, but having a companion might have helped immensely because Inyo would not have been alone when Terrill was absent. This might have curtailed the howling along with the other destructive behaviors Inyo exhibited when she was bored and alone. Inyo insisted on being an actual being with social needs, including a sense of proper limits.

This places Terrill firmly in the romantic philosophical tradition, along with Duman and others who fantasize about wolves and wolf-dogs and then find themselves disappointed by reality. In this tradition: "Ideals, ends, objectives are not to be discovered by intuition, by scientific means . . . by listening to experts or to authoritative persons; . . . *ideas are not to be discovered at all, they are to be invented,* not to be found, but to be generated . . . as art is generated" (Berlin 1999, 87). Given this type of thinking, it is not surprising that Inyo never had a chance (she was put down when she was three years old). Inyo was real; however, everything around her in Terrill's world was fantasy.

MISIDENTIFICATION

While we are discussing the issues that can develop because of false preconceptions that arise from romantic patterns of thought, Pierotti would like to address one of the most troubling cases he ever dealt with, one that involved either incompetence or deliberate misidentification by multiple individuals who considered themselves experts on wolf-dog iden- tification. This case took place in Minnesota in 1998 and 1999. A destruc- tion order for the animal in question was requested by Leslie Yoder, supervisor of Minneapolis Animal Control, and appealed in November 1998 (appeal was denied) before Pierotti was called into the case.

The particulars of this case were as follows: on October 14, 1998, animal control officer Cathy Johnson allegedly saw Luna, a large black dog, "jump over the fence of her owner's yard and entered the yard" at the owner's residence. As Officer Johnson "was attempting to catch and return Luna to her yard," she was bitten on the arm; the wound consisted of "two skin punctures and a laceration. The bite was ranked as minor" (City of Minneapolis 1998). According to the hearing document, the officer was attempting to return the dog to the yard *where she was already located.* Luna was seized and held at the animal control facility, where she was muzzled and locked in a four-foot by six-foot cage for the next two and a half months.

On October 20, after Luna had been held for a week, Peggy Callahan, executive director of the Wildlife Science Center in Forest Lake, Minnesota, was asked to examine Luna. She subsequently stated: "It is my professional opinion that the animal in question is part wolf. I base this on observing phenotypic expression of wolf characteristics, i.e., ear position and size, hair coat, leg length, leg position, tail shape, foot size, face configuration." Ms. Callahan, citing her thirteen years of "working professionally with wolves" averred that she "[made] this identification with confidence and experience." She then offered to house Luna in her facility in Forest Lake if she were given several thousand dollars per year (P. Callahan, letter to Leslie Yoder, October 22, 1998, cited from hearing records).

On October 23, Michael DonCarlos of the Minnesota Department of Natural Resources (DNR), who describes himself as a "Furbearer/Wildlife Depredation Specialist," wrote, "Based upon head characteristics, foot size, body proportions, behavior and other factors, I believe this animal is a wolf-dog cross. Although domestic dogs are highly variable in their physical and behavioral characteristics, especially when they are of mixed breeds there are some distinguishing characteristics that are usually only apparent in wolf-dog crosses. Because the animal I examined had many of these distinguishing characteristics, I'm quite certain the animal is a wolf-dog cross" (M. W. DonCarlos, letter to Leslie Yoder, October 23, 1998, cited from hearing records).

Ms. Yoder testified at the appeal that "she has been advised by the Minnesota Department of Health that there is currently no USDA approved rabies vaccination [for wolf-dogs] and that Luna must be considered to be not vaccinated for rabies" (J. B. Bender, letter to Leslie Yoder, November 10, 1998, cited from hearing records). Keep in mind that this was two years after the 1996 USDA advisory panel on which Pierotti participated had unanimously endorsed the idea that killed rabies vaccines should be recommended for use in all canids, both wild and domestic. Yoder further assumed that Luna was a wolf-dog rather than simply a dog (her owner had made sure Luna had her shots). It is important to acknowledge that Yoder was simply using language provided to her by Jeff Bender, state public health veterinarian, in a letter dated November 10, 1998. This reinforces three points made previously: (1) veterinarians are not scientists, (2) the USDA and the AVMA ignored scientific advice, backed in this by Dan Glickman, secretary of agriculture under President Clinton, and (3) this position is rooted in creationist rather than scientific thought, that is, that wolves and dogs are so different because of special creation that the same medical treatments cannot be used on both.

All of these issues became moot when I (Pierotti) had the opportunity to observe and examine Luna. I inspected the pedigree sent to the owner by the breeder, who described Luna as the descendant of a Groenendael

Igor de 3fi

FIGURE 10.2   Registered Belgian Groenendael. Note the high ears, long narrow snout, and long soft coat. (Photo provided by owner, who wishes to remain anonymous)

(a melanistic form of Belgian sheepdog) and a German shepherd, both AKC-recognized breeds (figure 10.2). Although the owner had presented this document at the hearing, it had not apparently carried any weight at the appeal, presumably because the "expert" opinions contended that Luna was a wolf-dog. When I arrived at the animal control facility, Supervisor Yoder did not want to allow me to examine Luna on the grounds that "she had already been evaluated by well-qualified experts." The attorney representing the owner quickly overcame this bluff, and the owner and I were eventually allowed to enter the facility and examine Luna. But Yoder refused to have Luna brought out of her cage because of "how dangerous she was," even when I offered to sign a waiver. The reason for their reluctance to have Luna inspected became apparent. Luna lay in her cage, emaciated and covered in her own fecal material. Yoder had determined that Luna was "so dangerous" that her cage would only be cleaned weekly, rather than daily as with other dogs.

It was obvious to me that Luna was in no way a wolf-dog. Despite the claims made by Callahan and DonCarlos, Luna had the long silky fur characteristic of a Groenendael, although initially this was hard to determine, given how badly caked with fecal material her fur was. The animal control staff refused to release Luna, so the attorney and I went to the latter's office, where I drafted a report describing the traits that identified Luna as a recognizable breed of dog, including her long and narrow muzzle, her slender small teeth, and her ears being close together atop her head in contrast to the well separated ears typical of wolves (ironically, Callahan's stationery featured a letterhead with a picture that clearly showed such wide-apart ears, yet she misidentified this trait in Luna). Of primary importance was Luna's long silky fur, totally unlike the pelage (not coat) of a wolf or high-percentage wolf-dog.

The attorney received an order from a judge allowing access. That afternoon the group of us (the owner, the attorney, and I) went back to the animal control facility, this time accompanied by a cameraman and reporter from a local television station. Yoder and her staff reluctantly allowed us access to Luna, but still would not allow her to leave her cage, although such permission has been routinely granted in every other dog case in which I have participated, even in Michigan. Luna's owner and I had to squeeze inside with her so I could examine her physical traits.

Her owner brushed Luna to remove the filth, while I pointed out to the reporter and cameraman how badly she was being mistreated, along with the features that showed she was in fact a member of an AKC-recognized breed of dog. Later, I drew the attention of the only animal control officer who behaved respectfully to an image of a Groenendael clearly shown on a poster in the office that depicted a couple hundred breeds of dog. I asked if the facility had ever dealt with a Belgian sheepdog of any kind, and the officer responded that they had not, at least in the three years that he had worked there.

The next day, based on our account (and the television coverage), a judge ordered Luna to be released to a veterinary clinic so her injuries and malnutrition from two and a half months of punitive captivity could be dealt with. One condition Yoder insisted upon before allowing Luna to leave is that neither the owner nor I file charges of animal cruelty against the animal control facility and its staff. When the owner and I went to see Luna just before her release, another animal control officer, Officer Johnson's boyfriend, attacked me and pushed me against the cages. One might have thought that these people, charged with protecting animal welfare, might have been pleased to see one of their charges released back to its owner, but it was apparent that Yoder and some of her staff wanted to see Luna killed. Their motivations were not clear, but it was obvious that in their minds Luna was a "dangerous animal" simply because they and their "experts" said so. No evidence would convince them otherwise, they were so strongly locked into the romantic mind-set that "*ideas* [or in this case the concept of biological identity] *are not to be discovered at all, they are to be invented,* not to be found, but to be generated . . . as art is generated."

As far as we know, Luna lived out her years with her owner, although they had to leave the city of Minneapolis, where Yoder and her subordinates would have continued to harass them.

CHAPTER ELEVEN

Living Well with Wolves and Dogs

OUR YUCHEE COLLEAGUE DAN WILDCAT once told a reporter, "Those who want to Dance with Wolves had first better learn to live with wolves" (Pierotti and Wildcat 2000, 1335). Humans who have good relationships with wolves and wolf-dogs are invariably those who respect the animals for what they are, and provide a secure and supportive environment in which these remarkable animals can thrive. As a counter to the previous chapter, here we present accounts of extraordinary individuals, both human and canid, who have achieved full and meaningful relationships.

We established in chapter 7 that Indigenous Americans lived quite comfortably with wolves. This does not mean that they entertained romantic fantasies; in fact, conflicts may have arisen between individual humans and wolves when relationships were being established, and this is an important aspect of the Cheyenne ritual traditions (Schlesier 1987). The key point is that most individuals of both species recognized that over the long term they both benefited from the relationship, although any member of either species who did not choose to participate was free to opt out and avoid interactions.

One individual of Euro-American heritage who was interested in establishing congenial relationships with wolves was the pioneering breeder of wolf-dogs, Gordon K. Smith. We elevate Smith to this level because it seems that no individual in American history, with the possible exception of some pre-contact Native Americans, has spent as much time interacting with wolves as Smith. Smith was born in northern Iowa around the beginning of the twentieth century, in a time and place when it was still possible to encounter wild wolves in a U.S. state that does not border Canada. His father showed him his first wolf when Smith was a young child and, as Smith puts it, "The seed was planted. Ever since that day I have experienced a sincere desire, bordering upon obsession, to gain a fuller knowledge of and to obtain a close relationship with old *Canis lupus*" (1978, 1). Smith spent at least fifty years studying wolves in the wild and living with an ever-changing number of captive wolves and wolf-dogs as he attempted to produce the ideal "wolf-a-dog," as he refers to them, an animal that looks like a wolf but is as easily socialized to interactions with humans as a dog: a truly domesticated wolf.

Smith provides a lengthy list of "a few things I have learned about wolves" in chapter 1 of his 1978 book, and most of these items fit closely with Pierotti's own observations and experiences. Smith's experience renders him an invaluable source of information, even if he was not trained as an animal behaviorist and his writing is not professionally edited. One key point he makes is that "bonding between strange [unfamiliar] wolves is sometimes very difficult, or even impossible" (7). This idea is relevant because it shows that the problem in bonding between humans and wolves is not *wildness* per se, as argued by Coppinger, Klinghammer, Duman, and Terrill, but rather familiarity. Humans who establish bonds with wolves are accepted as members of the wolves' social group. Smith provides ample evidence to support this point, observing, "Once intense psychological bonding has been established between wolf and wolf, or wolf and man, death or forcible separation are the only things that can break the bond." He concludes, "Wolves are friendly, gregarious creatures . . . and are very sad when forced to live alone. . . . They are quite willing to bond with any gentle, stable, warm-blooded creature, even man. . . . Wolves can be made happier in captivity than in the wild by an understanding and capable humanoid" (8).

Concerning situations similar to that Beth Duman described regarding her wolf, Nahanni, and its antagonism toward her husband, Smith stated: "Wolves seldom hold grudges; instant forgiveness is more often the case. However there are extenuating circumstances that can cause a wolf to hate for life. This hate, once established, cannot be broken. . . . [Otherwise,] wolves are normally peaceful creatures, sometimes breaking up fights and quarrels" (1978, 9–10). Pierotti's impressions are similar; wolves don't suddenly turn on owners if there is an established social bond. This suggests that Duman's husband did something to disrupt his relationship with Nahanni, as suggested by Duman's expanded comments in her presentation to the AVMA. Smith comments, "A wolf recognizes dominance or submission in humans, and if one is not inherently dominant enough, the wolf will not submit to [you]" (11). Here he refers to the process of establishing an initial relationship, not the dynamics of one that has already been instituted. Smith does not mean an animal that will not submit will attack, simply that it will resist being dominated. Interestingly, Smith notes, "Wolves do not trust alcoholics, drug abusers, or unstable temperaments . . . and can discern shifty treacherous people and respond by [retreating]" (11). Smith adds that he "turned down many would-be owners because my [wolf] male leaders did not approve of them" (12).

Smith's stated goal for breeding wolves and wolf-dogs was that he anticipated and feared the extermination of wild wolves and wanted to create what he describes as "Mankind's Wolf, just close enough to pure-bred as possible, [while retaining] the cooperation and domesticity of the dog" (1978, 13). The nature of "the dog" is somewhat mixed in Smith's mind, however; he conducted tests of fearfulness that showed "when the same state of fear was engendered [in domesticated dogs and pure wolves], the dog lost its cool much quicker than the wolf." Basically, he observed that when an unfamiliar human approaches, wolves will retreat into their enclosures as far back as they can, whereas most dogs, both large and small breeds, will "meet the strange human with fear, biting, howling" (14) and, on occasion, attack furiously. Alternatively, a well socialized animal might greet an unfamiliar human in a friendly manner. This was one major behavioral criterion Pierotti employed when assessing the identity of alleged wolves and wolf-dogs in cases in which he was involved.

Smith makes numerous insightful observations. He asserts that *Canis lupus nubilus*, the "buffalo wolf," the subspecies found on the Great Plains, was more "doglike"—that is, more easily socialized—when raised from a pup. This is interesting because *C. l. nubilus* would have been the subspecies with whom the Plains Indian tribes had their primary interactions. Given these congenial relationships (Fogg, Howe, and Pierotti

2015), this would fit well with Smith's assertions about sociability. Unfortunately, *C. l. nubilus* as well as any domesticated or socialized forms that may have descended from it are now presumed extinct.

Smith provides accounts of observing and interacting with wild wolves, concluding that if Europeans "could have tolerated the wolf, refraining from destruction of the species, it would have evolved into a state of semi-domesticity without much encouragement from [humans] and . . . could easily be interacting with *Homo sapiens* now, just like the dog" (1978, 122). From this it is clear that Smith, a Euro-American of his time, did not have a good understanding of the relationship between Indigenous Americans and wolves. He does, however, present the following account: "On May 21, 1935, I had gone to Canada . . . to pick up a couple of cubs [at] the home of an Indian family that owned a female house-raised wolf. . . . The first evening in my hosts' one room house, I noticed . . . the human toddler . . . lie down close to the wolf and play with the cubs. . . . Come supper time . . . the little toddler made for the old wolf, settled down flat on his [the toddler's] chest, with rump high in the air . . . proceeded to find a teat, then began nursing right along with the wolf cubs. . . . I was astounded at seeing the man child suck a she wolf. This brought about many gleeful chuckles from the entire family and my host said 'Wolf milk very good, do it all time. See how fat child is?' " (85). Smith describes the she-wolf cleaning the child after it defecated. He later recounts taking meat from a cow moose killed by wolves while the pack observed him. He howled with these wolves, recalling a statement made to him by another Indian, "Man is not until eat with wolf." He closes with, "That pack was my friend, I was their friend. Why couldn't this have been many times back in the dim primeval past?" (154).

Later on, in a critique of anti-wolf advocates, Smith states his position: "I will be the first to admit that a pure-bred wolf is a complicated structure, that few humans are ready and qualified to own one. But that is no reason . . . why a normal fairly intelligent person cannot set up lupus/humanoid relationships without conflict. . . . In all these years I have never had an adult wolf throw down the gauntlet to me. Sure they will test me; once in a while, lightly . . . to see if they can become pack leader. But open defiance, never" (1978, 156). Contrast this clear statement with the description provided by Duman concerning her husband's relationship with her household wolf.

Smith makes an interesting statement in discussing his breeding of wolves with dogs, which we have alluded to in previous chapters, "I know several AKC breeders who have bought 11/16 wolf-a-dogs . . . from us and replaced their registered German Shepherd studs with our [animal]" (1978. 161). These individuals apparently switched the papers of their former stud to the 11/16 crosses and sold the offspring as AKC-registered German shepherds. Presuming the truth of this account, which we see no reason to doubt, this raises a question of how many scholars studying "dog DNA" seem to miss repeatedly stated evidence of wolves being used to create German shepherd lines (see Frantz et al. 2016 for DNA evidence linking German shepherds to known wolf-dogs).

At the end of his book Smith provides instructions on how to breed a stable wolf-dog from scratch. (Smith refers to these animals as "hybrids" because in his time wolves and dogs were considered to be separate species. When he uses the term *hybrid* we substitute the more accurate *wolf-dog* in our accounts.) He starts with "a White German Shepherd male [crossed] with a female wolf of a large northern subspecies" (1978, 223).

PIEROTTI'S EXPERIENCE LIVING WITH WOLF-DOG CROSSES

When Pierotti began intensive work with wolf-dogs in 1989 at the University of New Mexico, he chose animals with the same pedigree described by Smith. The dam was a pure-bred, well-socialized Alaskan barren ground wolf (*Canis lupus tundrarum*) named Tundra, who weighed around thirty-five kilograms (seventy-five pounds) (figure 11.1). The sire was a white German shepherd (also around thirty-five kilograms) selected for a lack of hip dysplasia in his line. The animals Pierotti acquired were F1 (first-generation) crosses in genetic terms, meaning that dominant genes carried in the wolf line would be expressed and the recessive genes in the dog line were less likely to be expressed in the offspring's phenotypes.

I (Pierotti) became acquainted with Tundra while she was still pregnant and digging the whelping den. She was not aggressive in any way, although she did not want me to touch her, which was fine; I respected her caution and reticence. On the day of the birth (April 11, 1989), I selected two female pups and interacted with them for an extensive period so they would be used to my smell, sound, and touch even before their eyes and ears were functional. These individuals were chosen for two

FIGURE 11.1 Tundra, the dam of Pierotti's wolf-dog crosses. Note (1) the lack of stop (forehead), (2) the long, slender legs, (3) the deep but narrow chest, (4) the relatively short tail, (5) the head carried below the line from shoulders to hips, and (6) the relatively short but thick pelage, all of which are characteristics of wild wolves. (Photograph by R. Pierotti)

reasons: (1) I felt it would be easier to raise females than males, and (2) they did not have white markings, unlike their siblings, who tended to have white blazes on their rostra, which was a bit of their paternal genome showing up in facial markings. I visited the den every day to handle the puppies until it was time to remove these two individuals for hand rearing. During visits I noticed that the lighter-colored pup (Nuhmuh or Nimma) seemed to be the dominant puppy in the litter and was already attracting attention from her aunt, Keisha, who at two years of age was a University of New Mexico Lobo mascot (figure 11.2). Keisha harassed Nimma on a regular basis, following her around the yard. Any time Nimma showed any dominant or aggressive behavior, Keisha would roll her over and hold her down with a front paw. This was clearly dominance behavior directed toward a possible future competitor, strongly suggesting that Beth Duman's belief that one should never show dominance to a young wolf is incorrect. Here was clear, classic dominance behavior being directed at a two-week-old puppy by a full-grown female relative. Eventually, Nimma learned to escape Keisha's unwanted attention by slipping under a hog wire fence that divided portions of the yard under which she could fit, but Keisha could not follow.

FIGURE 11.2 Keisha's father was an F1 cross between a wolf and a domestic dog, making her 75 percent wolf. (Photograph by R. Pierotti)

When Nimma and her sister Tabananika were twenty days old, I removed them from their pack and took them to my house in the high desert surrounding Rio Rancho, New Mexico, where there was only one other house within a 100-meter radius. This house had a xeric backyard (sand and desert plants) with a covered patio, surrounded by a five-foot-high plank fence within which the puppies could be enclosed. The front yard was short grass surrounded by a four-foot cinder block wall except at the driveway, and contained a pine tree and cholla cactus. There was a resident pair of roadrunners (*Geococcyx californianus*) nesting on the roof next to the swamp cooler, and a striped skunk (*Mephitis mephitis*) lived underneath a shed just outside the backyard fence (who taught the rambunctious puppies that some animals around their size should be avoided).

I followed the protocol described by Gordon Smith for establishing strong social bonds: "Keep close contact with your cub until three or four months. . . . NEVER reject a wolf's desire to be close to you. This is its means of feeling secure" (1978, 88). The two puppies revealed distinct personalities but were very compatible with each other, with Nimma adopting the dominant role—she was on top in most wrestling matches and walked around with her small thin tail raised above her body line.

FIGURE 11.3 Tabananika at four weeks of age, showing the behavior that led to her name. (Photograph by R. Pierotti)

Taba was gentler and calmer. For the first couple of weeks, the puppies slept on a blanket in a cardboard box next to a futon on the floor on which I slept. Taba got her name (Tabananika, or sound of the sun) because every morning at first light, she would raise her head and give a puppy howl (figure 11.3), which was my signal to take them to the backyard to relieve themselves, after which we all returned to bed.

It was clearly beneficial to have two animals because they were each other's primary source of entertainment and social engagement, which meant they did not show any of the destructive behavior or aggression that Terrill describes as major components of her early interactions with Inyo (see chapter 10). The puppies were very "mouthy"—they would use their mouths to grab and explore things and would chew on hands and feet. If one ever actually bit down, we would quickly correct her by grab-bing her snout and saying, "No." The only incident of actual aggression toward a human occurred when Dr. Annett visited from Arkansas and had the interaction with Nimma described in chapter 9, which was similar in nature to the response Nimma had provoked from her aunt Keisha. Had Terrill dealt with Inyo in a similar manner rather than subscribing to the "no dominance paradigm" endorsed by Duman and

FIGURE 11.4 Nimma and Taba roughhousing at eight weeks of age. Nimma is showing dominance with raised tail and ears. Taba shows submissive postures with ears back and lowered head and tail. (Photograph by R. Pierotti)

other "wolf-dog experts" she consulted, she would have had far fewer difficulties and Inyo might still be alive.

As Taba and Nimma grew and matured, their personalities emerged more distinctly and they began to show increased activity and intraspecific agonistic behavior (Fentress 1967; see below). Nimma acted dominant toward Taba (figure 11.4) but showed uneasiness concerning new objects in her environment. Fentress reported neophobia in his hand-reared wolf at a similar age. In contrast, Taba quickly earned a reputation as our "wolf scientist" because she was interested in everything and would investigate anything new in her environment. She was the one who annoyed the skunk enough to cause it to spray (fortunately, the fence between them prevented serious impact). By the age of eight weeks Taba had figured out how to turn on the water spigot, which resulted in part of the backyard being flooded. I had to remove the handle. She also figured out how to get into cabinets with swinging doors. At times when I was looking for the puppies I could not find Taba—until I heard noises coming from inside the cupboard and opened the door, only to find Taba gazing out at me, perfectly calm. In contrast, Nimma was the one who initiated contact with new beings who might become part of the social group. Taba hung back from unfamiliar individuals of all species, and if Nimma would not approach an unfamiliar person, Taba would never go near that person. The sole exception was Dr. Annett; when she came to New Mexico for a visit, Taba immediately bonded with her.

As the puppies grew up and moved with my wife and me to Arkansas, a specific pack dynamic developed: Nimma assumed that I was her primary bond and Taba assumed a primary bond with Annett. The accepted social framework that emerged was that on group excursions I walked Nimma and Annett walked Taba. There was no serious competition, however, and the puppies were well socialized to both of us. They were not, however, incapable of aggression toward threatening individuals. During our move from New Mexico to Arkansas, when the pups were about fifteen weeks old, we drove at night because it was July and our car lacked air-conditioning. At a truck stop on I-40 in the middle of Oklahoma, Annett was walking Taba when a large unfamiliar man approached them. Sensing tension, the otherwise gentle, shy Taba raised her hackles and growled so impressively that the man backed off, apologizing.

We learned living with Taba and Nimma that the social bonds that developed between them and us were more intense, but more egalitarian in nature, than the bonds either of us had previously had with domestic dogs. The major reason seemed to be that wolves and wolf-dogs become fully adult as they mature, whereas the domestication process that produces dogs alters the timing of developmental events. Domestic dogs attain physical and reproductive maturity while still maintaining a juvenile-like relationship with their human companions. This is a primary reason that large domestic dogs like Molosser breeds can be so dangerous; they have the size and strength of an adult wolf combined with the temperament and social development of a half-grown wolf puppy. This also reveals the fallacy that many people embrace in their thinking about wolves and wolf-dogs: they assume that wolves are more dangerous because they are more independent, but that very independence makes them less dangerous—the only thing that makes wolves dangerous at all is their size and strength. Dangerous breed lists rank Molossers like presa Canarios and Caucasian ovcharkas because they are large and powerful, growing much larger than wild wolves, while retaining immature behavioral syndromes throughout their lives. These two breeds are much less common than wolf-dogs as companion animals; however, despite statements by "experts," such breeds, along with more common Molosser breeds such as boxers, pit bulls, and Rottweilers, are much more likely to attack and kill humans, especially children and small adults, than are wolves or wolf-dogs.

One consequence of differential timing of developmental events (neoteny) is that we can speak of "masters" when dealing with dogs because they retain juvenile behavioral characteristics throughout their lives. In contrast, by the age of two or three, wolves and wolf-dogs can have an "owner"—or, more accurately, favored companion—but they do not really want or need a "master," even if they accept dominance relationships (G. K. Smith 1978). Thus, shaping the behavior of the wolf or wolf-dog while young is crucial to the relationship that you will have with that animal throughout its life. Failing to recognize this is where Terrill went wrong in her relationship with Inyo, regardless of whether Inyo actually had any significant wolf ancestry. She failed to establish herself as the confident, trustworthy pack leader. As a consequence, as Inyo matured, she was unsure of her relationship with Terrill, who was emotionally unpredictable and seemed unsure whether she or Inyo should be the leader in any given situation. Because this crucial aspect of the relationship remained unresolved, Inyo and Terrill had regular conflicts, especially over food (see descriptions of their interactions in Terrill 2011).

In contrast, as we raised Taba and Nimma, we made it clear to them from the start that we were the dominant individuals in the relationship, as exemplified by Annett's eminently quotable "I'm the head bitch here, don't ever forget it." We fed them a combination of high-quality chow and raw bones and meat. During the crucial first three months, we established clear rules concerning interactions over food. (We have similar rules with our dogs today.) We established from the first that although we gave them bones, we could also take those bones away without conflict. This was achieved by offering a puppy a bone; if the puppy growled at us or lunged at the bone, we took it away, gently rolling the puppy while we growled at her. Consequently, even when they were full-grown adults, we could walk up to them, reach out a hand, and they would relinquish the bone. This was not something we did often, but it was firmly established that we *could* at any time we felt it was necessary. As a result, we lived almost conflict free with Taba and Nimma for their entire lives.

This is not to say that no conflicts ever arose within our group, but those that did take place were almost exclusively between Taba and Nimma themselves. These conflicts were predominantly low key and never went beyond posturing and growling; they involved play much more often than actual combat. Annett came up with a command, "Ears down," which we employed when we saw the beginnings of a conflict we thought might turn more serious. When wolves approach each other in a potential conflict situation, they hold their ears fully upright, whereas if at least one animal has its ears in a lowered position, it is indicating a submissive social position (see figure 11.4), which negates the potential for serious conflict. The command worked.

The theme of differential maturity and developmental processes is relevant to a frequently cited difference between dogs and wolves—that dogs pay much more attention to humans than do wolves (Hare et al. 2002; Miklósi 2007; Miklósi et al. 1998a, 1998b, 2007). This finding has been employed to support creationist-rooted arguments that, no matter what the "taxonomists" say, dogs and wolves are really "different species" (e.g., Coppinger and Coppinger 2001; Terrill 2011; Jans 2015). Upon reflection, however, it is obvious why dogs pay more attention to their human companions (Hare et al. 2002; Miklósi 2007); young children also pay much more attention to their parents than do adults. This does not mean that children and adults are different species, simply that they occupy different developmental stages. There is incredible ignorance in our society about evolution as a process, and it seems that many people assume an almost cookie-cutter similarity among individuals within a species, or among all species that are considered "wild," leading to typological thinking (for a contemporary example of such thinking, see Coppinger and Feinstein 2015). Such thinking ignores the importance of development (behavioral, morphological, and ecological) in evolution. Differences in developmental processes are responsible for the differences between wild and domestic forms of the same species (Ritvo 2010). Moreover, it is worth noting that these findings of Hare et al. and Miklósi et al. may be biased because they used pet dogs in their studies. Studies of hand-reared wolves showed that they were also responsive to human gestures, in some cases outperforming dogs, whereas kennel-raised dogs were much less responsive (Udell et al. 2008; Coppinger and Feinstein 2015).

Once they attained social maturity, Taba and Nimma established their own routines, and although it was obvious that they remained fond of their human companions throughout their lives, their strongest bonds were to one another and they lived their own lives. Once we moved to a rural area, Nimma rarely slept in the house; she stayed outside at night in almost all kinds of weather. Taba did sleep in the house, but on her own, whereas both had shared the bed with us when they were puppies and juveniles. They did appreciate air-conditioning on very hot days, but they also used their den, which was substantially cooler than ambient on hot days and substantially warmer on cold days.

Unlike the dogs we have now, Taba and Nimma did not bother to notify us about the presence of other wildlife, either inside or outside the enclosure. They would kill possums and raccoons without any fuss, and we often remained unaware they had done this until we found remains. They were very interested in the local coyotes, and this interest was returned in full, but they never called our attention to these interactions, although we heard them howling in response to coyote howls.

Despite their independence, they were highly socialized to their humans and formed a coherent social grouping with them. Included in this social group for most of their lives was a very pugnacious blue jay, *Cyanocitta cristata*. Shortly after we moved to Arkansas, someone brought a nestling blue jay into the zoology department. Dr. Annett, who is very good at raising birds, brought the nestling home and cared for her. As the jay matured and began testing her capacity for flight, I introduced the wolves to the fledgling jay by holding her in my hands and showing her to the puppies. Unexpectedly, Taba took the bird in her mouth; we were relieved when we opened Taba's mouth to find a chagrined but otherwise unharmed bird. After this drama, things settled into a routine, with Cool-J active in the air and the puppies active on the floor. Over the next eleven years, the different species developed interesting forms of cooperation. For example, neither the jay nor the wolf-dogs liked cats, and the jay had a specific alarm call she gave when she spotted a cat in the front yard. When she gave this call, the wolf-dogs would rush to the window where she was perched, and all three would direct general ill-will toward the offending feline.

We quickly realized that wolves are "sexist," preferring female humans over males in almost all situations. This was obvious from observing Taba's and Nimma's interactions with our students and colleagues. In fact, over their lives I was the only adult male human with whom Taba and Nimma had a friendly, congenial relationship because I was the only male who had been consistently in their lives since they were neonates. Even this bond could be tested by separation. When Taba and Nimma were sixteen months old, I went on sabbatical, leaving Taba and

Nimma in the care of Annett and a female graduate student who shared our large house in Arkansas. When I returned, Taba and Nimma had not seen me for fifteen weeks, and they initially greeted me with growls and avoidance, the typical response to an unfamiliar male. As soon as I said, "Girls, what's the matter?" however, they recognized my voice and went into submissive greeting behavior, lowering their ears and tails, whimpering and showing behavior typical of subordinate wolves greeting a senior member of the pack who has been absent for a while.

Taba and Nimma went through obedience training when they were six months old. Most striking about this experience was the difference in their responses. Training took place in a group, with around twenty other dogs and their human companions. Neither wolf-dog puppy showed any aggression to the other humans or dogs. Taba was more reticent and timid, especially when she was singled out for testing. In contrast, Nimma was the "star of the class," performing almost perfectly, except for one time when the instructor tried to take the lead from me to demonstrate a technique, at which point Nimma growled and raised her hackles. This was yet another example of dependence upon the bond. As long as she was performing with her human companion, Nimma behaved perfectly, but any attempt to interfere with that relationship was quickly challenged. As soon as I had the lead again, she settled down immediately.

These bonds with individuals and the established pack dynamics were extremely important. For example, we were told to work with the puppies individually: me with Nimma, Annett with Taba. When we took them on walks, we worked on various behaviors. Taba, however, did not want to be separated from Nimma, so she struggled more than she performed. In contrast, Nimma performed very well on her own with me. When the four of us worked together, Taba's performance was comparable to that of Nimma. Both wolf-dogs learned to "sit" and "stay," but were more reluctant to respond to "down," possibly because they interpreted this as being put in a submissive position.

Even temporary separations provoked strong emotional reactions from the wolf-dogs. One day when I was at home and Annett was at work, I noticed that Taba was vomiting and seemed distressed, so I decided to take her to the vet immediately. Knowing it would be difficult to handle both animals in such a situation, I called Annett to ask her to come home to be with Nimma. I then took Taba to the vet, leaving a very agitated

Nimma alone. Taba was fine; she had apparently eaten something bad and had mild food poisoning. When Annett arrived home, however, she was greeted by a very anxious Nimma, who flung herself into Annett's arms, whining, whimpering, and howling, and could not be comforted until I returned with Taba.

The depth of Taba's and Nimma's emotional bond was most obviously apparent whenever they had to be separated, even temporarily. As full-grown adults, each of them had to have minor surgery once in their lives. On these occasions, we took both of them to the veterinary clinic (despite the AVMA, our vet knew that the rabies vaccine works in all canids, and never hesitated to treat wolf-dogs or even pure wolves). Annett remained in the waiting room with whichever animal was not having surgery while I accompanied the animal undergoing the procedure. After surgery, the patient was returned to the waiting room, where she lay upon a blanket while recovering from the anesthesia. When Taba was recovering, Nimma stood over her protectively. The recovery lasted several hours, and periodically we had to take Nimma outside to relieve herself. Each time, Taba would lift her head and try to get up, and when Nimma returned to stand guard, Taba would relax again into semi-consciousness. When Taba finally recovered enough to go home, she wanted to relieve herself in our yard; Nimma walked beside her, attending her until Taba came into the house and lay down. Interestingly, when Nimma had her surgery, Taba did not stand guard over her sister during recovery but hid behind Annett, watching Nimma anxiously until she was back on her feet.

In our observations of their social dynamics, it seemed apparent that Nimma was the dominant one in the relationship, but such a simplistic concept cannot convey the subtleties in their mutual dependence. Although Nimma was socially more assertive and confident, she was smaller, weighing only seventy pounds to Taba's eighty. The size difference prevented Nimma from physically dominating Taba, rendering their relationship more equal. In addition, each had her own "sphere of dominance," an area where she took primary responsibility (Hand 1986; Pierotti, Annett, and Hand 1996). The idea of spheres of dominance is a theory developed by the brilliant animal behaviorist Judith Hand while she and Pierotti were studying male and female parental roles in western gulls, *Larus occidentalis,* in the 1970s. The prevailing wisdom was that male gulls were "dominant" over females within the pair-bond because they were larger and generally more aggressive (Pierotti 1981; Pierotti and Annett 1994). Yet Hand observed that there were numerous social conflicts in which females regularly controlled the outcome—that is, in situations where the female had more at stake than the male, such as behaviors relating to the nest and eggs. Dominance was not an all-or-nothing phenomenon but a variable situation that was resolved according to context (Hand 1986; Pierotti, Annett, and Hand 1996).

As mentioned above, Nimma was the "social secretary," the one who took the initiative in investigating and deciding which humans were safe to interact with. Taba was the "wolf scientist" and master of the inanimate—whenever we were in a new house or brought in a new piece of furniture, Nimma would show avoidance behavior and neophobia, whereas Taba would immediately investigate (this led to some comical adventures with new dog doors). For example, when we moved to Kansas in 1992 and Taba and Nimma experienced true winters for the first time, Taba became fascinated with the phenomenon of static electricity she observed during cold dry weather and would investigate various objects to see if a spark would pass between them and her nose. She would get a shock from the wood stove, then go to chair, a table, and even at times to us and place her nose next to these various objects at the same distance she had been from the stove, testing to see if a shock would result, and finally return to the stove to produce another spark—all within the space of a couple of minutes. In contrast, when Nimma had such an experience she would jump away, startled and agitated.

It is important to note that such complex and subtle behavior is a clear contradiction to the mechanical view of dogs (and by extension wolves) expressed in a recent scholarly work: "Dog lovers and researchers alike should be cautious about jumping to hasty conclusions about how much of behavior [in dogs] is actually associated with and guided by complex mental states. . . . We think the jury is still out on just how sentient or how 'smart' your dog might really be" (Coppinger and Feinstein 2015, 208–9). To see such obvious personality differences and different social roles in wolf-dog siblings born in the same litter, or to watch Taba testing different objects to see which produce static electricity, is a living refutation of such thinking. In their final chapter Coppinger and Feinstein even question whether or not dogs are "conscious," a question that was resolved some time ago by more sophisticated scholars of

FIGURE 11.5 Nimma and Taba next to the den they dug. (Photograph by C. A. Annett)

behavior like Jane Goodall and Marc Bekoff, and even by Hare and Woods (2013). It might be interesting to see Coppinger and Feinstein try to explain what Taba was doing in her repeated experiments with static electricity.

Around the time Taba and Nimma were achieving full maturity at about three years of age, we left our suburban lifestyle in Arkansas and moved to rural northeast Kansas, where we had seventy acres of mixed prairie, walnut and oak savannah, and bottomland woodlot. The house they shared with us was no longer a three bedroom on a half acre in town but a berm house (three sides covered by earth so that only the roof is aboveground) surrounded by a full acre enclosed by a six-foot-high chain-link fence. The enclosure was quite heterogeneous spatially, with a double row of pine trees on its north side serving as both a windbreak and a refuge (the gate through which unfamiliar humans entered was at the south side of the enclosure), and a single row of pine trees, which served to provide shade, on the southern side. There was also a tall mature white oak behind the house and a row of growing poplars. The animals Taba and Nimma interacted with were no longer neighborhood dogs and cats but coyotes, deer, wild turkeys, raccoons, and possums. They were able to go for long walks with us (on leads) without encountering other humans or their dogs, and without having to ride in the truck. This situation was clearly less stressful than living surrounded by other people, enduring regular and sometimes heavy traffic and the other sounds and annoyances associated with a suburban environment. Taba and Nimma dug a den, several feet deep, into the berm on the south side of the house (figure 11.5), into which both could disappear if they chose, especially on hot summer days.

FIGURE 11.6 Nimma in the snow. (Photograph by R. Pierotti)

Experiencing winter allowed them to manifest other behaviors we had not seen in New Mexico and Arkansas. They enjoyed snow and often went outside when it was snowing to engage in play behavior. Several times, calling their names, we were surprised to see a pile of snow suddenly stand up and shake itself off, revealing the wolf-dog sleeping beneath (figure 11.6). They also liked to use the roof, which, because it was a berm house, was only about eighteen inches above ground level in the back, as a vantage point from which they could survey the entire landscape. On occasion, they revealed their predatory side, taking raccoons, possums, squirrels, rabbits, and cotton rats that entered the enclosure. They charmed the local coyotes; male coyotes would come to the fence and court them. When the coyotes howled, Taba and Nimma would challenge them with their own howls. Taba had a low, mournful howl and Nimma went high. When their humans joined in, with Annett going higher than Nimma and me contributing to the low range with Taba, we could silence any coyote group within listening range.

Another interesting behavior shown by Taba and Nimma was their response to thunderstorms. It has been reported that "dogs will whine and cringe during a thunderstorm. . . . In rare cases their fear will develop to such an extent that that they will destroy furniture, claw at windows until they bleed, or defecate. . . . In severe cases . . . dogs may develop an illness or have a fatal heart attack" (Hare and Woods 2013, 230). In contrast, Nimma was defiant toward thunderstorms, and she recognized that lightning precedes thunder. Our location provides an expansive view toward the west, the direction from which thunderstorms come in Kansas. When Nimma saw lightning she would rush to the western side of the enclosure and run along the fence line; the sound she made was not a growl or howl but a full-throated roar, a behavior she also showed toward unfamiliar and unexpected humans or dogs. Taba would join her in running the fence line, but she was usually less vocal. This response was directed toward thunder and lightning rather than rain, which the wolf-dogs seemed to regard as a distracting and sometimes annoying event.

Taba and Nimma always had full run of the house, whether or not we were at home. This had also been true in Arkansas. We did not leave food out, and all clothing had to be put in chests or closets, because items left lying about might disappear into their den, where they apparently achieved totemic status. Taba was particularly fond of one of Dr. Annett's caps, which vanished one winter and was not seen again until the following summer, when it was found carefully cached under the pine trees at the north end of the enclosure. Despite such minor issues, the wolf-dogs never did any damage to furniture or the house. The one thing that seemed to trouble them was that a berm house has only one entrance (and exit)—they seemed to think we needed a back door as we had had in Arkansas, through which they could retreat when things became stressful. Except on hot, humid nights Nimma typically slept outside on sentry duty, while Taba slept in our bedroom, sometimes on the bed, especially when one of us was away traveling, but more often on the floor next to the bed.

Another remarkable skill that Taba and Nimma displayed was the ability to recognize human family members. Although they were typically wary of unfamiliar humans, this was not the situation with immediate family members. When members of our families visited, Taba and Nimma initially reacted with caution, but after sniffing them, they very quickly switched to acceptance and treated family members the way they did other pack members—seeking attention and affection and giving it in return. There was one exception to this rule, and in this case Nimma provided a warning we should have heeded. Recalling G. K. Smith's (1978) description of his animals' aversion to alcoholics, drug users, and individuals with psychological problems, there was one family member toward whom Nimma did not respond in a friendly way; she even showed mild hostility, growling and moving away when this person approached. Years later, after Taba and Nimma were gone, this individual revealed mental health problems.

Gordon Smith, the master breeder and scholar of wolves, argued that difficult as captivity could be for wolves, in many ways living wild was worse: regular hunger, combined with the danger and harassment directed at wild wolves by humans, created a stressful environment that often made them fearful and insecure. He felt that the ideal situation for wolves was living with humans—not in a cage but in a secure relationship where their needs were met through their interaction with humans or, as he describes it, "men hunting in groups, flanked by wolves in a common pack" (1978, 191), which is the basic model we described in the introduction. We tried to provide something similar for Taba and Nimma. They were "with" us, but they did not "belong" to us. They were individuals, with a remarkable amount of control over their lives within a secure environment.

We were astounded by how little food they needed; they had a couple of meaty bones daily, supplemented by dog biscuits, and there was always a bowl of good-quality kibble available; only Nimma, however, seemed to eat from the bowl, and even she would not empty it for several days. Their general consumption rate was so low, in fact, that it took us a while to realize that Nimma was eating very little in late spring 2000. We took her to the vet, where it was determined that she was suffering from congestive heart failure. At the age of eleven years and two months, she passed on June 11, 2000, attended by her pack.

The surviving threesome had one last howl that evening, but after Nimma was gone Taba would never howl again, even though she survived Nimma by more than four years. After Nimma's death, Taba was very depressed. She had lost her lifetime companion; they had never been apart for more than a few hours over the previous eleven years. Her complex emotional state further refutes the mechanistic thinking of Coppinger and Feinstein (2015). We provided Taba with canid antidepressants and made sure that she was not left alone at all for several months after Nimma's death.

As Taba adjusted to being the only nonhuman in the pack, she came up with her own unique solutions to being alone. While Nimma was alive, the wolf-dogs had experienced a somewhat fractious relationship with our donkeys and buckskin quarter horse, who sometimes grazed alongside the fence surrounding the enclosure. Nimma was fond of charging the fence at the equid group, with Taba providing semi-enthusiastic backup, both enjoying the reaction as the equids scattered in flight, sometimes jumping in the air when they were startled. A few months after Nimma was gone, however, I saw our old horse approach the fence while Taba stood on the other side. The horse lowered his head, pressing his nose against the fence, and we were surprised to see Taba lean forward and lick him on the nose. Apparently she had established a truce, because a few weeks later we found Taba sleeping against the fence, while on the other side was our old male Sicilian donkey, his body pressed against her through the fence. Again, it would be interesting to see Coppinger and Feinstein try to explain such behavior within their machine-model-driven ideas of how canids lack complex mental states.

After Nimma passed, Taba went on long cross-country walks with me, now establishing friendly interactions with domestic dogs in the surrounding countryside who had seemed intimidated when all four of us used to walk together. On her own, Taba became a sweet-natured, nonaggressive, highly social canid. She still was shy and uneasy around unfamiliar humans, but otherwise she lived a very peaceful four final years. Her passing was less dramatic than Nimma's; one warm August afternoon when Taba was fifteen years and four months old, I went out to read under the oak tree behind the house. Taba walked out with me, lay down at my feet, and never woke up.

We learned from living with wolf-dogs that if one makes sure the animals understand what is and is not acceptable behavior, and if one takes care not to tempt animals left on their own, humans and wolf-dogs can share perfectly good and peaceful lives together with only minimal conflicts. In fact, the conflicts we had with Taba and Nimma were much less intense and less frequent than we have with our current pack, consisting of two border collies and a Tennessee Walker hound. The hound, who joined us on her own as a full-grown adult, had to be rolled on multiple occasions early on when she climbed on tables to try to eat off plates when guests encouraged her. The hound and our older border collie, who came to us as a stray when she was a year old, have engaged in two serious fights that we had to break up, with Annett dragging the hound away and me tackling the border collie and holding her on the ground. They have now established a very amiable relationship, with little conflict. The second, younger border collie is more independent and less well socialized so far, at the age of thirty-two months, than either Taba or Nimma were at the same age. All in all, living with wolf-dogs is no more stressful than living with any other large dog breed, as long as you train them properly and avoid foolish mistakes.

One final memory of Taba and Nimma is the lesson they taught me when they were about a year old. Working on a study of wild canids in Arkansas, my students and I had been preparing skulls by removing flesh so the skulls could be cleaned as specimens. We were left with a large bowl of meat scraps removed from the remains of trapped coyotes. I took this bowl home and offered it to Taba and Nimma, who sniffed the contents and reacted with the most apparent disgust I have ever seen in canids. They looked at me as if I had offended them deeply. This experience convinced me that much more was going on with these nonhumans than I had imagined, because they enthusiastically ate any other meat they were brought, such as beaver. They showed me that they clearly had defined boundaries they would not cross.

SEREN AND PETER: WOLVES AS AMBASSADORS

Neither Taba nor Nimma was well enough socialized to unfamiliar people to be used in educational outreach about wolves. Fortunately, in 1992, working an early wolf case in Kansas City, Pierotti encountered a couple who worked in wolf and wolf-dog rescues who were successful in outreach: Dr. Noreen Overeem, a veterinarian with a specific interest in wolves, and her partner, Tracy McCarty, a former member of the military who has had extensive experience in working with captive wolves around the United States. They ran an outreach program for local schools and public parks that introduced people to socialized wolves. When Pierotti began to work with them, Overeem and McCarty were in the process of switching their primary outreach animal from an aging female wolf to her grandson, eleven-month-old Peter, who was highly socializeable and well trained.

One of the initial outreach events we did together was a full assembly at a suburban high school in eastern Kansas. Pierotti was speaking on the stage while holding Peter's lead. This was Peter's first appearance before a large number of strangers and, like many in such a situation, he was nervous, so he regurgitated his breakfast on the stage. It is easy to imagine the effect this had on an audience of teenage *Homo sapiens*. The teachers who had invited us were running to get mops, buckets, and paper towels when Peter's grandmother, Seren, handled by McCarty, walked over and ate all of the regurgitate off the stage. If the cries of "Eeeew, gross!" had been loud before, they now increased by an order of magnitude. Pierotti waited a moment and then pointed out to the students (and teachers) that this was perfectly normal wolf behavior, and that one principle of wolf life is that food should not be wasted. This quickly led to a discussion of families and why what we call wolf "packs" are actually nuclear families. We also established how similar wolves were to humans in a lot of their behavior and attitudes.

The next time Pierotti worked with Peter and Seren was one of Seren's last public appearances. She was getting old and beginning to develop the cancer that would eventually take her life. We wanted this to be an occasion, so we set up a scenario during a class session of Pierotti and Annett's course Principles of Ecology in the context of a discussion of trophic cascades and the role of wolves in shaping ecosystems. The class met in a large lecture hall, designed to accommodate around 200 students, which was dim but not dark so students could easily see slides. At about the forty-minute mark of the seventy-five-minute class (the point where students usually begin to nod off, especially in a warm, semi-dark room), McCarty and Overeem brought in Peter and Seren, each woman walking a wolf down one aisle from the back of the room toward the podium. As the students became aware of who was passing by them we could hear the murmurs increase to a general low rumble. When they reached the podium where Pierotti was standing, he turned up the lights and introduced the guests, both human and canid, to the class. At this point, as at the high school, we had a wide-ranging discussion and question-and-answer session about wolves and their relationships with humans. Many of the students were surprised to see wolves behaving so calmly around humans, and many of them wanted to touch the wolves. Fortunately, Peter and Seren were used to this, and everyone who wanted to was able to see them close up and touch them. There was so much interest in the wolves, in actually experiencing something rather than just

FIGURE 11.7 Peter (*left*) and Seren (*right*): wolves in the classroom. (R. Pierotti's personal collection)

hearing about it that we had difficulty clearing the classroom to allow the next class to start (figure 11.7).

Over the next several years Peter, McCarty, and Pierotti conducted regular outreach programs at dozens of schools, ranging from day care centers to universities; at public parks; and even at nature-themed stores around eastern Kansas. (We could not take Peter into Missouri because it is against the law to have a wolf in that state, even for purposes of education.) Pierotti developed a continually growing respect for Peter and his calm dignity under circumstances that even most humans would find trying. Peter clearly felt regular social stress—most of the time he had his tail down and at times even between his legs, which is a signal of unease in *Canis lupus*. Yet never once did he show any aggression; he never growled or even curled his lip, no matter how many well-intentioned strange *Homo sapiens* pawed at him or wanted to gaze into his eyes. Peter was exceptional in his ability to discern personalities among his fans. On one occasion he spent almost an hour with his head resting in the lap of a young girl who had serious family problems and was very stressed. On another occasion, at an elementary school in Lawrence, Peter spotted an eight-year-old boy with cerebral palsy in the midst of about fifty students. The wolf led McCarty right up to the child, looked into his eyes, and sniffed him gently. The attention made the boy feel special—and interestingly, being singled out by Peter seemed to change the boy's classmates' attitudes toward him; teachers told Pierotti afterward that fellow students who had previously avoided the boy started engaging with him much more after Peter identified him as special.

The most moving experience we had with Peter was his relationship with a teenager in Topeka who was blind and autistic. McCarty and Pierotti had been invited to bring Peter to participate in Earth Day activities. During the noon break, when we were alone with Peter, the boy's teacher brought this student in to meet him. As soon as the student touched Peter, and Peter sniffed him gently in response, all tension seemed to disappear from the boy, and he sat next to Peter, gently stroking him. Even after the bell rang, marking the end of the break, the boy insisted on staying with Peter. His teacher, who was so thrilled to see the boy engage emotionally and socially, allowed him to accompany Peter on the rest of his activities throughout the day. When we returned the following fall to the same high school, we found the boy waiting. His teacher told us that the boy had not wanted to get on the bus to come to school that morning and had been in a bad mood until the teacher told him, "You know Peter is going to be at school today." The boy immediately became cooperative and even eager to get to school to see his friend.

Pierotti has done hundreds of outreach events over his career as an educator, but never has he seen students as receptive or engaged as when Peter was present. Pierotti introduced Peter as "the greatest gentleman you will ever meet" but asked students to be respectful, to consider how they would feel if they "found themselves facing a room full of unfamiliar wolves." Getting the students to think of it that way made them careful in their behavior; if anyone started to get rambunctious, all we or their teacher needed to say was, "You don't want to upset Peter," and even the most boisterous young boys would settle down. The reason this was so effective was it was very clear to the students how much respect McCarty and Pierotti had for Peter, both in the way we spoke about him and in the way the three members of two different species interacted. These experiences with Peter became one of the principal motivating factors for writing this book, because Pierotti, like the students, knew that with Peter he was in the presence of dignity and grace (figure 11.8).

FIGURE 11.8   Peter waiting patiently to meet yet another group of students. Note how far apart his ears are in this relaxed pose. (R. Pierotti's personal collection)

One irony we never discussed with students was that the dynamic between Pierotti and Peter was very different in public situations than when Peter was at home in his enclosure with other wolves. Like other wolves, Peter was sexist, and aside from his outreach activities with Pierotti he did not interact with adult male humans. The social group in which he lived consisted entirely of women. When Peter was stressed in classrooms, he often leaned against Pierotti for reassurance, yet when he was in the world he controlled, Peter had no interest in engaging with Pierotti. When Pierotti entered the enclosure, Peter retreated and lowered his ears and tail, showing avoidance behavior typical of a wolf encountering an unfamiliar human. Peter showed no aggression, yet it was clear to Pierotti that even though Peter regarded him as trustworthy and part of his group in the outreach dynamic, on his home ground with the other wolves in his pack, he wanted no part of Pierotti.

Pierotti recognized that, as with Nimma and Taba, Peter's behavior resulted from a "sphere of dominance" dynamic. In outreach situations, where Pierotti assumed the role of confident male, speaking to and engaging the audience, and Peter showed at least some submissive behavior, Peter was willing to accept the role of subordinate and even to

FIGURE 11.9 Peter (*foreground*) and his younger cousin 4×4 in their shared home enclosure. Note that although Peter is not overtly displaying dominance, 4×4 is showing an obviously submissive posture, with lowered tail and body. (R. Pierotti's personal collection)

rely on Pierotti for emotional support in stressful situations. On his own turf, however, where Peter was the dominant individual in his pack, which consisted of a younger female cousin and two younger male cousins, Peter had control (figure 11.9). The arrival of Pierotti, another potentially dominant male, even though he was familiar, threw Peter into conflict. He did not wish to appear subordinate to Pierotti, which might have reduced his status in the eyes of his younger, but larger, male cousins. Peter chose to react with avoidance, a behavior quickly copied by the other members of the pack. He lost no status, and therefore avoided conflict with a human male with whom he had a social relationship independent of his pack. Being the perfect gentleman, he chose to act uneasy rather than offending Pierotti by trying to appear dominant.

Thinking about this situation allowed Pierotti to understand how wild canids can maintain relationships with humans while also being effective members of a family group within their own species. Peter was not in a conflicted situation when navigating between the female humans who were his owners and handlers and his status in his pack (unlike his relationship with Pierotti as a male handler). This dynamic is one reason that in the scenario with which we open this book, we describe the initial relationship between wolf and human as one established between women and a female wolf, with the male humans working with the wolves as fellow hunters but not necessarily maintaining the stronger, more subtle relationships that existed between women and wolves. This situation is mirrored in stories from Plains tribes of North America: it is women (sometimes children) who are guarded by and provided for by wolves (see Blackfoot, Cheyenne, and Lakota stories in chapter 7). In contrast, in the story of the old Blackfoot man who offers meat to the wolf and the wolf makes him a better hunter in return, there is no indication that the two have an extended relationship in which the human is cared for by the wolf. Similarly, in the Tsistsista (Cheyenne) creation story, Maiyun (the male protector wolf spirit) chose to teach the newcomers, the humans, to hunt on the grasslands (Schlesier 1987), but this again suggests a cooperative but not intimate social relationship.

LUPEY, THE WOLF OF CAMBRIDGE

In the mid-1960s John Fentress, a doctoral student at Cambridge University, raised a male Canadian timber wolf from the age of four weeks, at which point the puppy had a mass of 3.2 kilograms (about 7 pounds) (Fentress 1967). At this point, as with Taba and Nimma, the puppy's ears were flopped over, his eyes were deep blue (like Nimma's), and he had a short snout typical of all canid puppies of a similar age. Fentress chose to raise Lupey with what he called "firm cooperation" (340)—patience and relaxed contact as opposed to constant applications of "strict dominance." Fentress studied the development of animal social behavior and kept careful records of Lupey's progress, supported by photographs and 3,000 feet of Kodachrome motion film for comparison with domestic dog behavior.

Fentress noted than within the first twenty-four hours of living with him, Lupey changed from avoiding humans to approaching them, which shows that young animals are capable of quickly bonding (at least at a superficial level) with those who provide food. For the first two days, Lupey had to be force fed, which was not an issue with Taba and Nimma; however, Pierotti had been interacting with his puppies since the day they were born. Like Taba, but not Nimma, Lupey howled when he was left alone at night, but would stop if he was brought into bed (341). Four weeks later, at eight weeks of age, the pup became more active and aggressive. This is the age Inyo was when Terrill began having difficulty with her because Terrill failed to establish an effective social relationship. Lupey also showed aggression issues around this time—he growled when Fentress tried to remove a mop he was playing with, snapped when restrained, and at one point drew blood when lunging after a piece of meat offered to him. Around this age Nimma had struggled when restrained on a lead but did not snap, and Taba never struggled or snapped. This was probably because both had already been trained not to growl or snap when food was offered, and Pierotti took bones or toys away from them as part of their training. In fact, any lunging or snapping meant the food would be withdrawn until they calmed down and behaved in a proper calm, submissive fashion.

By the age of ten weeks, Lupey became more restless and needed lots of exercise. He escaped the room he was kept in by climbing onto a desk and out the attic window; however, he would not jump off the roof and so returned to his room. Fentress wisely decided to leave the city and move to the country. At this age Taba and Nimma were also highly active, but they always had access to the yard through a dog door and we regularly walked them on leads.

By the age of thirteen weeks Lupey was chained to a doghouse on a six-meter (twenty-foot) chain because he was restless and destructive when confined to the house. This is never a good idea. Being chained places severe stress on any canid, even domestic dogs. Lupey killed several chickens at this age. He did not eat the chickens, but after each kill he ate double his normal rations, indicating that the predatory act had some impact on his appetite. Interestingly, Lupey never showed aggression toward domestic dogs, instead soliciting play behavior.

At twenty weeks Lupey began to treat strange objects and unfamiliar humans with caution. This is normal behavior; at this age wolf (or dog) pups begin to be more independent of their mother, so their caution is adaptive—pups are beginning to explore their environment on their own and should be cautious around unfamiliar individuals of any species. This is also around the age when we took Taba and Nimma to obedience training, where they also showed no aggression toward domestic dogs. Around this time Lupey also developed a calmer personality within his social group: his play with familiar humans became gentler and he learned to sit and shake hands.

As Lupey grew and matured, he showed increased independence but remained gentle with humans and dogs. He was allowed to run free with farm dogs, which reduced his energy and restlessness. At the age of ten months, Lupey was shipped to the United States, accompanied by Fentress. Once he slipped his collar and escaped. He was found and held by a young girl whom he had never seen before, which showed that he was well socialized to humans.

During his third year Lupey grew even calmer and more friendly toward humans. During a blizzard he escaped from his pen for two days and was found six miles away, playing with dogs and children. Like all wolves and high-percentage crosses, he enjoyed being near familiar humans, and although he was cautious around unfamiliar ones, he showed no aggression. Lupey remained in close social contact with humans for more than three years, actively seeking attention from those with whom he had an intimate relationship. The only time he snapped as an adult was when he was receiving medical treatment.

WOLVES AND THE PIANIST

The counterexample to Ceiridwen Terrill and her failure to live successfully with Inyo is the case of Hélène Grimaud, a young woman who was able to establish special rapport with wolves. Grimaud, born in Aix-en-Provence of a French father and an Italian (Corsican) mother, reveals a mind-set every bit as romantic as Terrill's, but with a crucial difference. As a concert pianist, she has been forced to develop a personal discipline that serves her well in her chosen areas of interest, which include wolf conservation. This allowed her to establish a better-reasoned approach to her own desire to live with wolves.

She describes her first physical experience with a wolf, which took place late at night when she was walking a friend's German shepherd in Tallahassee, Florida, where she was living at the time. "And it was at that instant that I saw it for the very first time. The silhouette of a dog—but with one glance, and despite the dark—one could instantly tell that it was not a dog. The animal had an indescribable walk—tense, furtive, as if was making its way through a tunnel that was barely large enough. Its eyes had an almost supernatural glow. They gave off a muted light, violet and wild. . . . The creature looked at me, and a shiver ran through me—neither fear nor anxiety, just a shiver" (Grimaud 2003, 203).

The mystery animal was accompanied by a Vietnam veteran, about whom she had been warned; he is "not right in the head" and "dangerous."

The man informed her, "It's a she-wolf," and warned her not to move because, "she's shy." Now Grimaud had a mystical experience that changed her life, giving her a new purpose and direction: "Slowly she approached me. . . . She came up to my left hand and sniffed it. I stretched out my fingers, and all by herself, she slid her head and then her shoulders under my palm. I felt a shooting spark, a shock, which ran through my entire body. The single point of contact radiated throughout my arm and chest, and filled me with gentleness . . . a most compelling gentleness, which awakened in me a mysterious singing, the call of an unknown, primeval force. At the same moment, the wolf seemed to soften, and she lay down on her side. She offered me her belly" (2003, 205). The man told her that he had never seen his animal do that before. "It's a sign of recognition and trust, even a sign of submission" (206).

After this experience, Grimaud knew that she wanted to share her life with wolves. She represents a good counterpoint to Ceiridwen Terrill; unlike Terrill, Grimaud did not rush in; she developed a plan before taking on the rigors of living with pure wolves. With USDA approval, Grimaud got all the necessary permits, acquired land in a rural area well away from cities, and had the terrain altered and secure fences built, with the help of volunteers. She established a preserve, the Wolf Conservation Center, funded by her income as a classical pianist. By 2003, she had four pure wolves living in her preserve.

Today Grimaud's Wolf Conservation Center is a well-established facility housing gray wolves, red wolves, and Mexican wolves. Hélène Grimaud has shown that you can have a romantic view of wolves and still be a successful and active companion to the animals, making sure that they have contented, fulfilling lives; you just have to have a solid plan and take the animals' lives as seriously as your own.

Conclusion

THE FRIENDLY PREDATOR

THE MAJOR ENIGMA OF the process we refer to as the first domestication is that wolves and dogs can seem so affectionate and willing, even driven, to create strong and persistent social bonds with humans that it becomes easy for humans to become overconfident and even careless about these four-legged relatives that share our lives so easily. Even in their fully domestic state, however, dogs remain predators, highly evolved carnivores that know how to kill. Given opportunities, they may kill other entities that humans value—chickens, cats, on occasion sheep and cattle, and even humans themselves. As wolves, even the smallest, gentlest dogs remain predators, regardless of how much they have changed physically and behaviorally from their wolf ancestors.

For example, a major factor in the extinction of the thylacine in Tasmania was that it was blamed for sheep killings that were in fact carried out by domestic dogs (D. Owen 2003). Pierotti once talked with a neighbor who was convinced that coyotes were killing his poultry until he found that his own German shepherd had been the culprit; our local coyotes don't seem to kill livestock of any kind. A large, powerful dog that is highly aggressive, starved, or feels threatened can become dangerous, especially to small children and frail elder humans. Such dogs are far more deadly in terms of human health than wolves have ever been, as reflected in the mortality statistics (Sacks et al. 2000).

As long as humans considered themselves as fellow predators of wolves, we lived comfortably with them (Pierotti 2011a). This was also true of the barely domesticated wolves that were the original Russian laiki, dingoes, or American Indian dogs. They were our boon companions, sharing our hunts and kills, living with us in a reciprocal arrangement. Once we began controlling the lives of other species, like goats, sheep, horses, pigs, and cattle, that had previously been our shared prey, however, we felt the need to try to change wolves as well. This can be seen most clearly in the Maremma, Kuvasz, and other breeds reared to live their lives with sheep: now they guard these ungulates from wolves or coyotes rather then help us hunt them down (Coppinger and Feinstein 2015).

Sometime during the Middle Ages Europeans turned on wolves and began to treat them as creatures of Satan, a being who exists only in the minds of people of European ancestry (Coates 1998; Coleman 2004; Grimaud 2003; Pastoureau 2007; Pierotti 2011a). The reasons humans create imagined villains are complex, and often involve the Christian Church, which worked hard to break relationships between humans and carnivores during the period referred to as the "Dark Ages" (Coates 1998; Pastoureau 2007). In our view, however, the real darkness came later, with the massive slaughter of both people and wildlife perpetrated by colonial Europeans as they spread themselves around the world from the fifteenth through nineteenth centuries (Sale 1991; Coates 1998; Mann 2002, 2005; Coleman 2004). Underlying this change in human attitudes was the (apparently conscious) decision to recast humans, especially followers of Christianity, from being predators to being prey (Coates 1998; Pierotti 2011a), best exemplified in the image of Christians as "the flock of a loving and protective shepherd," notwithstanding the fact that real shepherds are actually the most relentless predators on their flocks (Pierotti 2011a).

As we have described, in most parts of the world and for almost the entirety of human history, humans interacted with wolves, or wild dogs, in a primarily positive manner. We have described Japanese okami, regarded as important allies for farmers because they helped keep their fields safe from deer and wild pigs, and the wild wolves of Hokkaido, invited to become friends and hunting companions of the Ainu people. We have discussed the loneliness of Australian Aboriginal peoples, who were so excited when dingoes joined them 5,000 years ago that they remade their creation stories to include the only other placental mammal to join them in a land dominated by strange pouched and egg-laying mammals incapable of being proper companions for active, imaginative humans (Rose 2000, 2011). We have presented a wide range of traditional stories from Indigenous American peoples describing a range of positive relationships with wolves beginning in the mists of prehistory and continuing up until at least the later part of the nineteenth century (Fogg, Howe, and Pierotti 2015).

In each case, these conditions prevailed until Europeans or Euro-Americans arrived, bringing fear, hatred, and livestock. The wolves or dingoes then became the targets of relentless persecution (Coleman 2004; Walker 2005; Rose 2011). There is a story among Native Americans that the first action taken by Europeans upon arrival was to establish a bounty on wolves; the second was to place a bounty on the Indians themselves (Coleman 2004; Pierotti 2011a). In both situations, acceptable evidence of a kill was the scalp because heads were too heavy to transport in large numbers. In North America and Australia, genocidal war targeted both wolves and their Indigenous human companions. This was also the fate of the Ainu and their wolf companions on Hokkaido (Walker 2005). These are not rumors or guesswork: the wars and the processes involved are well documented (McIntyre 1995; Coleman 2004; Walker 2005; Rose 2011).

Only in the southern islands of Japan, where disease did not clear a path for their invasions, did the British and Americans wipe out only wolves, leaving the Indigenous population in place (Walker 2005). Another exception to this pattern was southern Siberia, where extreme climate, combined with immunity to the diseases that devastated American and Australian Indigenous people, allowed Indigenous peoples to survive (Sale 1991; Pierotti 2004, 2006, 2011a, 2011b; Rose 2000, 2011; Mann 2005). Only in the last few centuries did European Russians move effectively into Siberia, and it was not until the Soviet Union period from 1920 to 1990 that Siberian Indigenous peoples and wolves really suffered persecution, after enduring oppression during tsarist times because of the punitive requirements imposed on them for fur trapping (Lincoln 1994). Siberia remains a challenge, and there is something in the Russian spirit that allowed them to embrace the wolflike laiki, if not the wolves themselves (Cherkassov 1962; Voilochnikov and Voilochnikov

1982; Forsyth 1992; Beregovoy 2001, 2012, Beregovoy and Porter 2001), allowing this to be one European culture where humans remain comfortable with very wolflike dogs.

We opened this book with a story about a young female wolf joining a group of humans. We would like to close by providing some understanding concerning how many times during the short history of modern humans they found wolves who did not fear them. Most were young males and females looking to establish their own social group who decided that these strange, sparsely haired primates who walked on their hind legs might make decent pack members. The wolves were much better adapted than humans to the harsh ecological conditions of Europe, Asia, and North America. They became our companions, our teachers, our guides. Some became the models for the wolf Maiyun and his female companion, who taught the proto-Tsistsista people how to hunt; others became the solitary wolves who guided and protected Lakota and Blackfoot women who found themselves alone and in need of food and companionship; yet another guided the survivors of the Sand Creek massacre.

We have previously alluded to the case of Romeo, a large, young black wolf who spent seven years interacting in a friendly manner with hundreds of people and many more dogs in the outskirts of Juneau, the capital city of Alaska (Jans 2015). Pierotti has heard numerous accounts from people working in "wild areas" about "friendly wolves" who approach humans out in the bush, sometimes seeking food but more often simply wanting to play or experience a mutually respectful friendly interaction. The example of Romeo, however, is particularly well documented and is worth exploring in some detail to relate it to many of the themes we have developed.

Nick Jans, the author of *A Wolf Called Romeo*, is an experienced woodsman and naturalist who has lived in Alaska and worked with wildlife for more than thirty years. The story he tells sounds extraordinary to Western-indoctrinated ears; however, he is simply describing one of the more recent examples of a dynamic that has played out many thousands of times over the millennia, especially in North America, where wolves are not hostile to humans because they spent thousands of years interacting with them as equals. The late Gordon Haber, the most experienced and dedicated of Alaskan wolf scientists, who worked for more than forty years with Denali's wolves, described a species that shows no aggression toward humans, is mostly curious about human activities, even when people intrude in wolf areas, and is defensive only when puppies and dens are threatened (Haber and Holleman 2013). Jans struggles at times in telling the story of Romeo, which actually strengthens his account—he is not simply a naïve wolf enthusiast or advocate who wants to merge minds with the wolf; he is an experienced naturalist who wants to provide a measured account of a profound experience.

Jans seems conflicted about the idea of Romeo and what he represents. As an Alaskan naturalist, Jans is very much a hunter and has killed wolves while working with Native Alaskans, a fact to which he alludes several times. Encountering Romeo throws him off balance: "The true measure of distance between wolves and dogs lies in the eyes. A dog's may display intelligence and engagement, but being caught in a wolf's unblinking gaze is like standing in the path of a laser. . . . This black wolf's deep amber irises hold all that force, but something more radiated from him that I'd never sensed in any other wild wolf: a relaxed acceptance of my presence" (2015, 5). Jans recognizes that Romeo is unlike other wolves he has encountered, but perhaps more important here is that Jans seems convinced that wolves and dogs are different species. He keeps trying to place Romeo in the category of "Other," part of a world separate from the human. The Other is an all-encompassing concept implying a variety of meanings, including its position as a widely accepted synonym for difference—that is, not of Western human society (Fabian 1990). Romeo keeps confounding Jans, revealing that some wolves behave in a way he cannot quite comprehend.

In contrast, one individual who makes quick contact with Romeo and establishes solid rapport is Dakotah, Jans's female retriever. The book's cover shows Romeo and Dakotah greeting each other. It seems clear that Dakotah recognizes one of her own kind, regardless of what her master thinks about them being different. A running theme in Jans's book is that Romeo is drawn more to dogs than to humans, and the feeling seems to be generally mutual. In particular, Jans describes Romeo as having especially close relationships with Dakotah, one of his neighbor's border collies, and a large black Lab mix named Brittain, the companion of one of the book's most interesting human characters, Harry Robinson. All of these dogs are female; as a dispersing young male, Romeo may be seeking a partner, a mother figure, or perhaps a combination of both.

Romeo becomes a fixture of the Juneau scene, especially during the period from the freezing of the lake in the fall until it melts in the spring. Most humans respond well to his presence, although it is clear that there are some who would like to see him killed simply for existing. Interestingly, however, although this latter group seems to threaten and posture a lot, no local in the small and well-integrated community of Juneau ever takes action against Romeo, although the reader gets the impression that many are simply waiting for the wolf to commit a transgression that would justify their feelings.

Romeo, though playful with most dogs, at times administers discipline to those who cross a line. Jans has three dogs: Dakotah; Gus, an older male black Lab and former seeing-eye dog; and Chase, a yearling blue heeler (Australian cattle dog). Jans describes an encounter between Chase and Romeo:

> I had Chase's leash firmly under my boot—or so I thought.
> A sudden unexpected tug and off she went, a snarling blur
> flying straight for the wolf. . . . The wolf picked up the charge
> and came bounding to meet her full on. . . . The two met in
> an explosion of snow, the wolf wide jawed, bounding, paws
> slamming down to pin our dog. In that heart-rattling instant,
> Chase disappeared under the wolf. . . . I'd screwed up in a way
> I could never forgive myself.
>
> Then a blue gray shape exploded out of the snow, headed
> back as fast as she'd gone in, yelping all the way. Lips pulled
> back in a grin . . . the wolf bounded along behind her a few
> feet, then trailed back. . . . Though she was quivering and
> her fur stiff with frozen saliva, we went over every inch and
> couldn't find the least ding or bruise. (2015, 34)

Rather than injure, or even kill, the impolite dog, all Romeo did was show appropriate wolf discipline toward an insubordinate junior.

On at least two occasions when Pierotti was walking Nimma she administered similar quick and impressive discipline to small barking dogs that charged her, and our senior border collie regularly administers similar discipline to our border collie puppy when she transgresses. Such behavior is highly ritualized, and even though it looks (and sounds) deadly at first, injuries almost never result.

Later on, playing with an Akita puppy belonging to a local veterinarian, Romeo "seized the puppy by the neck and bounded off into the willows" (2015, 131). The vet, filled with fear and guilt, plunges into the willows after them, when suddenly "his puppy came scampering towards him, whining. . . . A nose to tail exam under [the vet's] practiced hands couldn't detect a single laceration or bruise" (132). The key message to take from such encounters is that wolves, and their domestic descendants, can discipline misbehaving young group members in a way that may suggest serious danger to humans, yet the canids involved recognize that what is going on is part of normal group dynamics. Those chastened may be highly chagrined, their pride injured, but any resulting scars are at worst psychological.

The most intriguing human in Romeo's world is Harry Robinson, who appoints himself as one of Romeo's translators to the general public (we wish Robinson would publish his own account). Jans (and others) refer to Harry as the "Wolf Whisperer," a cliché that has probably outlived its usefulness. Robinson refers to Romeo as his "friend," which seems a more appropriate term and suggests a more egalitarian relationship. Robinson seems to understand better than others that the presence of Romeo is a gift, one that should be honored rather than exploited. When Robinson and his dog Brittain go on extended excursions throughout the local landscape, if Robinson "howls," Romeo often will join them: "The three would trace game trails and visit . . . the rolling spruce-hemlock forest" (Jans 2015, 158). Robinson witnesses Romeo taking beaver as prey and finds evidence that the wolf has taken goat-antelopes (*Oreamnos americanus*), which are common in the mountains above Juneau. Romeo is clearly an accomplished hunter, not dependent on being fed by humans (or scavenging). Romeo also protects his friends: "Romeo once sensed something ahead on one of their walks, bristled, then lunged forward, growling, as a locally known brown [grizzly] bear and grown cub appeared around a bend in the trail, a few dozen feet away. As the wolf charged in defense of his pack, the bear turned tail, and Romeo completed the rout" (159). Through such actions, Romeo reveals that he is, in fact, a fully functioning wild wolf, capable of challenging any other creature in his environment. Thus, when he chooses to associate with human and canid companions, and even to protect them, he also shows where his priorities lie.

In our view, the significant aspect is not so much that Robinson is a "Wolf Whisperer" but that Romeo may be something even more impressive, a "Human Whisperer." While we are on this theme, it seems relevant to discuss one of Cesar Millan's greatest insights—his assessment of some of the happiest, most emotionally stable dogs in America:

> I think that dogs that live with homeless people often have the most fulfilling, balanced lives. . . . These dogs don't exactly look like AKC champs, but they're almost always well behaved and nonaggressive. Watch a homeless person walking with a dog and you will witness a good example of pack leader–pack follower body language. . . . The dog follows either beside the human or just behind her. . . . Dogs don't know the difference between organic and regular dog food, they don't think about groomers, and in nature, there aren't any vets. . . . Homeless people . . . walk from place to place, pick up cans, and seek a meal and a warm place to sleep. This lifestyle might be unacceptable to many humans, but for a dog this is the ideal, natural routine that nature created for him. He is getting the consistent amount of primal exercise that he needs. . . . He is free to travel. . . . Exploration is a natural animal trait. . . . Balance in a dog's life isn't created by giving them material things . . . [but] by allowing them to fully express the physical and psychological parts of their being. Living with a homeless person, a dog migrates [and] . . . works for food. (Millan and Peltier 2006, 130–31)

Robinson and Brittain provide Romeo with something very similar; they wander like a wolf group, sometimes finding food and sometimes adventure. As Millan points out, "Homeless owners aren't pampering their dogs . . . although the dogs can sense that their owners are happy about having them around" (Millan and Peltier 2006, 202–3). Homeless owners provide "someone to follow, who'll eventually lead the dogs to food and water and a place to rest" (203). One gets the sense that had any of the humans around Juneau been willing, Romeo would have been pleased to have a constant companion, happy to take the lead if that was what the human wanted, as Robinson seems to have done in the proper context.

Over time Romeo learned to respond to communication from Robinson. "He would obey a number of my commands, although he would carefully consider them beforehand. . . . He'd observe a situation and reason it through . . . but he definitely knew what the word *no!* meant" (Jans 2015, 158). On one occasion when Romeo grabs a pug in his mouth, Robinson shouts, "*No!*" and the wolf releases the dog. In another case, Robinson intervenes in and diffuses a physical confrontation between Romeo and a large husky mix.

Over the years Romeo seems to have won over the majority of humans he encountered through behaving with dignity and friendliness. Reading Jans's and Robinson's accounts, Pierotti was reminded of his time spent watching Peter interact with other humans, and how impressed he was that the most dignified and best behaved individual around was the nonhuman, and that he had the ability to change the behavior of the humans themselves.

Romeo lived to the age of at least nine before he was needlessly shot and killed by a couple of young men from outside the Juneau area; Jans provides clear photos of "the killers," as he refers to them. Peter lived to the age of fourteen and died of old age, safe and well cared for in his enclosure. Some might compare the fates of these two extraordinary individuals as supporting evidence for Gordon Smith's (1978) argument that the ideal situation for wolves in the modern world is living with humans— not in a cage but in a secure relationship where their needs are met through their interaction with humans. Romeo lived free, but he died violently. Was this a better or worse life than the safe, secure existence that Peter experienced, cared for by loving humans, never in danger?

Such relationships as Romeo developed with people have happened many times, in many places. It continues to happen up to the present in locations like Juneau, and it will continue to happen in the future if humans are prepared to receive the gift. The wolf called Romeo offered such a gift. Some, like Harry Robinson, were ready to receive it. Others, like Nick Jans, knew there was a gift of being offered but were unsure about how to respond, influenced by the fear-filled legacy the Christian Church has impressed upon people of European heritage over the last thousand years (Coleman 2004; Pastoureau 2007; Pierotti 2011a).

The exchange of such gifts between human and wolf has changed both species over the last forty or so millennia. Some of the wolves changed physically, becoming more neotenic, and came to be considered dogs. There is a tendency today to view dogs as our best nonhuman friends and wolves as our implacable enemies, but this way of thinking fails to recognize that these different physical forms remain in essence the same being, with the same predatory nature. We fear wolves and love dogs, although dogs kill far more people. Because we consider dogs under our "control," we do not desire their extermination the way many seem to do with wolves, although some efforts at breed-specific legislation are exceptions to this general rule. There is much talk in modern America about other cultures "fearing our freedom," but it is contemporary Americans who fear freedom, especially the kind they see in the wolf, and they use this fear to justify hatred and killing.

It is time to recognize that we define dogs behaviorally as much as physically. As we discussed in chapter 4, this topic might create controversy in archaeological circles or within standard systematics and evolutionary thinking, where DNA and anatomy are the primary traits considered, because behavior and ecology cannot be effectively incorporated into phylogenies. Defining wolves behaviorally rather than anatomically is, however, more true to the process used in the domestication of wolves.

Our contemporary struggle to recognize the wolves within our dogs emphasizes the differences between Western and Indigenous ways of understanding. Indigenous peoples saw the canids who hunted with them, guarded their villages, shared their lives, and at times saved them from the persecution of both Europeans and other tribes (Marshall 1995; Fogg, Howe, and Pierotti 2015) as wolves, or in Australia, dingoes (Rose 2000, 2011), whereas to Europeans, canids living with humans must be "dogs," regardless of their independence or ability to switch from living with humans to living on their own. An example of this dichotomy in worldviews: when we were publishing an article on the relationships between Indigenous Americans and wolves (Fogg, Howe, and Pierotti 2015), our coauthor Nimachia Howe showed a draft of our manuscript to several Blackfoot elders. Their response to our supposed major finding—that there existed strong positive relationships between tribes and wolves—was instructive, best summed up as "What's the big deal? Doesn't everyone know this?"

Despite problems and harassment, native peoples still try to stand by their old lupine friends. Recent efforts by the U.S. federal government to remove protection from wolves have generated strong opposition from tribes. Chippewa tribes in Wisconsin, Minnesota, and Michigan requested the Wisconsin Department of Natural Resources to prohibit the killing of wolves in the ceded territory of northern Wisconsin. Jim Zorn, executive director of the Great Lakes Indian Fish and Wildlife Commission, said, "Tribes believe the Wisconsin hunt is biologically reckless and would be culturally harmful to Chippewa, for whom wolves are culturally important. . . . 'How should we sanction the killing of our brother?' The Voigt Intertribal Task Force of the Great Lakes Indian Fish and Wildlife Commission passed a motion unanimously opposing the killing of [wolves] and claiming all wolves in the Wisconsin ceded territory as a necessary prerequisite to a population that would fully effectuate the Tribes' rights" (Knight 2012). None of these tribes were consulted before these measures were instituted, and all tribes in Wisconsin have requested that no wolf hunts take place and no hunters be allowed to kill wolves on tribal lands (Lewis 2013; Fogg, Howe, and Pierotti 2015).

Some tribes have gone well beyond simple requests, employing serious leverage. In retaliation for wolf hunts, six bands of Chippewa in northern Wisconsin declared their intention to spear a near-record number of walleyes during the annual spring harvest, terminating a 1997 agreement with the state and effectively shutting down the sport fishing season (Fogg, Howe, and Pierotti 2015). Such actions resulted because the relationship between the state and tribes has become increasingly strained, primarily because the tribes strongly opposed opening a wolf hunting and trapping season starting in 2012 (Knight 2012; Lewis 2013). Wolves feature prominently in origin stories and legends of the Chippewa; they are associated with courage, strength, and loyalty. Chippewa bands living in Wisconsin were allotted several "slots" of the total wolves allowed to be killed—but they refused to kill a single wolf (Lewis 2013; Fogg, Howe, and Pierotti 2015). Despite 500 years of colonization, some Indigenous Americans are still trying to protect wolves, their companions, teachers, and creators.

To an Indigenous person Romeo would simply have been what he was: a very sociable wolf, capable of living with people or without them, but preferring to share his life with them. He would have made an excellent hunting partner and might have shared their fires on cold nights, providing an alert presence that would have allowed them to sleep safely.

In contrast, the Western worldview ties itself into knots over the concept of a wild wolf that is friendly toward human beings, even though such entities were probably an everyday experience, at least in North America, until the arrival of Europeans. The real tragedy of Romeo's life was not that he was killed by a couple of ignorant fools with guns instead of brains, but that the gift he offered was taken up by so few.

Allen, D. L. 1979. *Wolves of Minong: Isle Royale's Wild Community.* Ann Arbor: University of Michigan Press.

Allen, P. G. 1986. *The Sacred Hoop: Recovering the Feminine in American Indian Traditions.* Boston: Beacon.

Allman, J. M. 1999. *Evolving Brains.* New York: W. H. Freeman.

Altmann, J. 1990. "Primate Males Go Where the Females Are." *Animal Behaviour* 39:193–95.

American Veterinary Medical Association. 2001. "A Community Approach to Dog Bite Prevention: American Veterinary Medical Association Task Force on Canine Aggression and Human-Canine Interactions." *Journal of the American Veterinary Medical Association* 218 (11): 1732–49.

Anderson, E. N. 1996. *Ecologies of the Heart: Emotion, Belief, and the Environment.* New York: Oxford University Press.

Anderson, T. N., B. M. vonHoldt, S. I. Candille, M. Musiani, C. Greco, D. R. Stahler, D. W. Smith, B. Padhukasahasram, E. Randi, J. A. Leonard, C. D. Bustamante, E. A. Ostrander, H. Tang, R. K. Wayne, and G. S. Barsh. 2009. "Molecular and Evolutionary History of Melanism in North American Gray Wolves." *Science* 323 (5919): 1339–43.

Animal Behaviour Society. 2012. "Guidelines for the Treatment of Animals in Behavioural Research and Teaching." *Animal Behaviour* 83:301–9.

Annett, C. A., and R. Pierotti. 1999. "Long-term Reproductive Output and Recruitment in Western Gulls: Consequences of Alternate Foraging Tactics." *Ecology* 80:288–97.

Annett, C. A., R. Pierotti, and J. R. Baylis. 1999. "Male and Female Parental Roles in a Biparental Cichlid, *Tilapia mariae.*" *Environmental Biology of Fishes* 54:283–93.

Atleo, E. R. (Umeek). 2004. *Tsawalk: A Nuu-chah-nulth Worldview*. Vancouver: University of British Columbia Press.

Audubon, M. R. 1960. *Audubon and His Journals*. New York: Dover. Originally published in 1897.

Ballinger, F. 2004. *Living Sideways: Tricksters in American Indian Oral Traditions*. Norman: University of Oklahoma Press.

Barsh, R. L. 1997. "Forests, Indigenous Peoples, and Biodiversity." *Global Biodiversity (Canadian Museum of Nature)* 7 (2): 20–24.

———. 2000. "Taking Indigenous Science Seriously." In *Biodiversity in Canada: Ecology, Ideas, and Action,* edited by S. A. Bocking, 152–73. Toronto: Broadview.

———. n.d. "Nonagricultural Peoples and Captive Wildlife: Implications for Ecology and Evolution." Unpublished MS.

Basedow, Herbert. 1925. *The Australian Aboriginal*. Adelaide: F. W. Preece and Sons.

Basso, K. 1996. *Wisdom Sits in Places*. Albuquerque: University of New Mexico Press.

Bastien, B. 2004. *Blackfoot Ways of Knowing: The Worldview of the Sijsikaitsitapi*. Calgary: University of Calgary Press.

Bazaliiskiy, V. I., and N. A. Savelyev. 2003. "The Wolf of Baikal: The Lokomotiv Early Neolitic Cemetery in Siberia." *Antiquity* 77:20–30.

Beeland, T. D. 2013. *The Secret World of Red Wolves: The Fight to Save North America's Other Wolf*. Chapel Hill: University of North Carolina Press.

Bekoff, M. 2001. Review of *Dogs: A Startling New Understanding of Canine Origin, Behavior, and Evolution,* by Raymond Coppinger and Lorna Coppinger. *The Bark: Dog Is My Co-pilot*. http://thebark.com/content/dogs-startling-new-understanding-canine-origin-behavior-and-evolution.

Bell, G. 1982. *The Masterpiece of Nature: The Evolution and Genetics of Sexuality*. Berkeley: University of California Press.

Belyaev, D. 1979. "Destabilizing Selection as a Factor in Domestication." *Journal of Heredity* 70:301–8.

Belyaev, D. K., and L. N. Trut. 1982. "Accelerating Evolution." *Science in the U.S.S.R* 5:24–29, 60–64.

Benton-Benai, E. 1979. *The Mishomis Book: The Voice of the Ojibway*. St. Paul: Indian Country.

Berard, H., and K. W. Luckert. 1984. *Navajo Coyote Tales: The Curly Tó Aheedliinii Version*. Lincoln: University of Nebraska Press.

Beregovoy, V. 2001. *Hunting Laika Breeds of Russia*. Bristol, TN: Crystal Dreams.

———. 2012. "The Concept of an Aboriginal Dog Breed." *Primitive and Aboriginal Dog Society Newsletter* 34:5–12.

Beregovoy, V. H., and J. Moore Porter. 2001. *Primitive Breeds—Perfect Dogs*. Arvada, CO: Hoflin.

Berk, A., and C. D. Anderson. 2008. *Coyote Speaks: Wonders of the Native American World*. New York: Abrams for Young Readers.

Berlin, I. 1999. *The Roots of Romanticism*. Princeton: Princeton University Press.

Bettelheim, B. 1959. "Feral Children and Autistic Children." *American Journal of Sociology* 64 (5): 455–67.

Binford, L. R. 1980. "Willow Smoke and Dogs' Tails: Hunter-Gatherer Settlement Systems and Archaeological Site Formation." *American Antiquity* 45:4–20.

Bocherens, H., D. G. Drucker, M. Germonpré, M. Lázničková-Galetová, Y. I. Naito, C. Wissing, J. Bruzek, and M. Oliva. 2014. "Reconstruction of the Gravettian Food-Web at Predmosti I Using Multi-isotopic Tracking (13C, 15N, 34S) of Bone Collagen." *Quaternary International* 359–60:261–79. http://dx .doi.org/10.1016/j.quaint.2014.09.044.

Boesch, C. 1994. "Cooperative Hunting in Wild Chimpanzees." *Animal Behaviour* 48:653–67.

Boudadi-Maligne, M., and G. Escarguel. 2014. "A Biometric Re-evaluation of Recent Claims for Early Upper Paleolithic Wolf Domestication in Eurasia." *Journal of Archaeological Science* 45:80–89.

Brackenridge, H. M. 1904. *Journal of a Voyage Up the River Missouri Performed in 1811*. Edited by R. G. Thwaites. Cleveland: A. C. Clark.

Brenner, M. 1998. "A Witch among the Navajo." *Gnosis* 48:37–43.

Bright, W. 1993. *A Coyote Reader*. Berkeley: University of California Press.

Bringhurst, R. 2008. *The Tree of Meaning: Language, Mind, and Ecology*. Berkeley: Counterpoint.

Brody, H. 2000. *The Other Side of Eden: Hunters, Farmers and the Shaping of the World*. Vancouver: Douglas and McIntyre.

Brown, A. K. 1993. "Looking through the Glass Darkly: The Editorialized Mourning Dove." In *New Voices in Native American Literary Criticism*, edited by A. Krupat, 274–90. Washington, DC: Smithsonian Institution Press.

Browne, J. 2003. *Charles Darwin: A Biography*. Vol. 2, *The Power of Place*. Princeton: Princeton University Press.

Bruchac, J. 2003. *Our Stories Remembered: American Indian History, Culture, and Values through Storytelling*. Golden, CO: Fulcrum.

Bshary, R., A. Hohner, K. Ait-el-Djoudi, and H. Fricke. 2006. "Interspecific Communicative and Coordinated Hunting between Groupers and Giant Moray Eels in the Red Sea." *PLoS Biol* 4 (12): e431. doi:10.1371/journal. pbio.0040431.

Buller, G. 1983. "Comanche and Coyote, the Culture Maker." In *Smoothing the Ground*, edited by B. Swann, 245–58. Berkeley: University of California Press.

Cann R. L., M. Stoneking, and A. C. Wilson 1987. "Mitochondrial DNA and Human Evolution." *Nature* 325 (6099): 31–36.

Catlin, G. 1973. *Letters and Notes on the Manners, Customs, and Conditions of the North American Indians: Written during Eight Years' Travel*. New York: Dover. Originally published in 1842.

Cherkassov, A. A. 1962. "Notes of Hunter-Naturalist" [in Russian]. *Academy of Sciences of the USSR, Moscow*.

City of Minneapolis. 1998. Minutes of administrative hearing for animal control case (#98–9952). November 17.

Clark, P. U., A. S. Dyke, et al. 2009. "The Last Glacial Maximum." *Science* 325:710–14.

Clode, D. 2002. *Killers in Eden*. Sydney: Allen and Unwin.

Clutton-Brock, J. 1981. *Domesticated Animals from Early Times*. Austin: University of Texas Press.

———. 1984. "Dog." In *Evolution of Domesticated Animals*, edited by I. L. Mason, 198–211. New York: Longman.

———. 1995. "Origins of the Dog: Domestic and Early History." In *The Domestic Dog*, edited by J. Serpell, 8–20. Cambridge: Cambridge University Press.

Clutton-Brock, J., et al. 1977. "Man-made dogs." *Science* 197:1340–42.

Clutton-Brock, T. H., and P. H. Harvey. 1977. "Primate Ecology and Social Organization." *Journal of Zoology, London* 183:1–39.

Coates, P. 1998. *Nature: Western Attitudes since Ancient Times*. Berkeley: University of California Press.

Coleman, J. T. 2004. *Vicious: Wolves and Men in America*. New Haven: Yale University Press.

Colshorn, C., and T. Colshorn. 1854. *Märchen und Sagen*. Hannover: Verlag von Carl Rümpler.

Coppinger, R., and L. Coppinger. 2001. *Dogs: A Startling New Understanding of Canine Origin, Behavior and Evolution*. Chicago: University of Chicago Press.

Coppinger, R., and M. Feinstein. 2015. *How Dogs Work*. Chicago: University of Chicago Press.

Coppinger, R., L. Spector, and L. Miller. 2010. "What, if Anything, Is a Wolf?" In *The World of Wolves: New Perspectives on Ecology, Behavior, and Management*. Calgary: University of Calgary Press.

Corbett, L. K. 1995. *The Dingo in Australia and Asia*. Ithaca: Cornell University Press.

Coren, S. 2006. *The Intelligence of Dogs: A Guide to the Thoughts, Emotions, and Inner Lives of Our Canine Companions*. New York: Free Press.

———. 2012. *Do Dogs Dream? Nearly Everything Your Dog Wants You to Know*. New York: Norton.

Corless, H. 1990. *The Weiser Indians: Shoshoni Peacemakers*. Salt Lake City: University of Utah Press.

Crockett, C., and C. H. Janson. 2000. "Infanticide in Red Howlers: Female Group Size, Male Membership, and a Possible Link to Folivory." In *Infanticide by Males and Its Implications*, edited by C. P. van Schaik and C. H. Johnson, 75–98. Cambridge: Cambridge University Press.

Crockford, S. J. 2006. *Rhythms of Life: Thyroid Hormone and the Origin of Species*. Victoria, BC: Trafford.

Crockford, S. J., and Y. V. Kuzmin. 2012. "Comments on Germonpré et al., *Journal of Archaeological Science* 36, 2009, 'Fossil Dogs and Wolves from Paleolithic Sites in Belgium, the Ukraine and Russia: Osteometry, Ancient DNA and Stable Isotopes,' and Germonpré, Lázničková-Galetová, and Sablin, *Journal of Archaeological Science* 39, 2012, 'Paleolithic Dog Skulls at the Gravettian Předmostí Site, the Czech Republic.'" *Journal of Archaeological Science* 39:2797–801.

Crook, J. H., and J. C. Gartlan. 1966. "Evolution of Primate Societies." *Nature* 210:1200–1203.

Czaplicka, M. A. 1914. *Aboriginal Siberia: A Study in Social Anthropology.* Oxford: Clarendon.

Darwin, C. 1859. *The Origin of Species.* London: J. Murray.

Dawkins, R. 2006. *The Selfish Gene.* Oxford: Oxford University Press.

———. 2015. *A Brief Candle in the Dark: My Life in Science.* New York: HarperCollins.

Dayton, L. 2003a. "On the Trail of the First Dingo." *Science* 302:555–56.

———. 2003b. "Tracing the Road Down Under." *Science* 302:555.

Dean, W. R. J., W. R. Siegfried, and A. W. MacDonald. 1990. "The Fallacy, Fact, and Fate of Guiding Behavior in the Greater Honeyguide." *Conservation Biology* 4:99–101.

Deloria, V., Jr. 1992. "The Spatial Problem of History." In *God Is Red,* 114–34. Golden, CO: North American Press.

Derr, M. 1995. "*Happy People* and Laikas." http://retrieverman.net/tag/happy-people-a-year-in-the-taiga/.

———. 2011. *How the Dog Became the Dog: From Wolves to Our Best Friends.* New York: Overlook/Duckworth.

Descola, P. 2013. *Beyond Nature and Culture.* Chicago: University of Chicago Press.

Diamond, J. 1992. *The Third Chimpanzee: The Evolution and Future of the Human Animal.* New York: HarperCollins.

Dillehay, T. D., C. Ocampo, J. Saavedra, A. O. Sawakuchi, R. M. Vega, M. Pino, et al. 2015. "New Archaeological Evidence for an Early Human Presence at Monte Verde, Chile." *PLoS ONE* 10 (11): e0141923. doi:10.1371/journal.pone.0141923.

Dobie, J. F. 1961. *The Voice of the Coyote.* Lincoln: University of Nebraska Press.

Dombrosky, J., and S. Wolverton. 2014. "TNR and Conservation on a University Campus: A Political Ecological Perspective." *PeerJ* 2: e312. doi:10.7717/peerj.312.

Drake, A. G., M. Coquerelle, and G. Colombeau. 2015. "3D Morphometric Analysis of Fossil Canid Skulls Contradicts the Suggested Domestication of Dogs during the Late Paleolithic." *Nature Scientific Reports* 5 (8299): 1–8. doi:10.1038/srep08299.

Druzhkova, A. S., O. Thalmann, V. A. Trifonov, J. A. Leonard, N. V. Vorobieva, N. D. Ovodov, A. A. Graphodatsky, and R. K. Wayne. 2013. "Ancient DNA Analysis Affirms the Canid from Altai as a Primitive Dog." *PLoS ONE* 8 (3): e57754.

Dugatkin, L. 1997. *Cooperation among Animals: An Evolutionary Perspective.* Oxford: Oxford University Press.

Dulik, M. C., et al. 2012. "Mitochondrial DNA and Y Chromosome Variation Provides Evidence for a Recent Common Ancestry between Native Americans and Indigenous Altaians." *American Journal of Historical Geography* 90:229–46. doi:10.1016/j.ajhg.2011.12.014.

Duman, E. 1994. "Is It a Wolf-Hybrid? What Will It Do? Raising a Wolf-Hybrid and Housing Requirements." Paper presented at the 131st American Veterinary Medical Association Annual Meeting, San Francisco, July 9–13.

Dunbar, R. I. M. 1988. *Primate Social Systems*. Ithaca: Cornell University Press.

———. 1992. "Neocortex Size as a Constraint on Group Size in Primates." *Journal of Human Evolution* 20:469–93.

———. 1995. "Neocortex Size and Group Size in Primates: A Test of the Hypothesis." *Journal of Human Evolution* 28:287–96.

———. 1998. "The Social Brain Hypothesis." *Evolutionary Anthropology* 6:178–90.

———. 2000. "Male Mating Strategies: A Modeling Approach." In *Primate Males*, edited by P. Kappeler et al., 259–68. Cambridge: Cambridge University Press.

Edwards, A. 2013. "Bear Strikes Up Unlikely Friendship with a Wolf as Photographer Captures Both Animals Sharing Dinner on Several Nights." *Daily Mail*, October 4. http://www.dailymail.co.uk/news/article-2443974/Bear-WOLFs-unlikely-friendship-caught-camera-photographer.html#ixzz3tDbmgIW.

Eisenberg, J. F., N. A. Muckenhirn, and R. Rudran. 1972. "The Relation between Ecology and Social Structure in Primates." *Science* 176:863–74.

Eldredge, N., and S. J. Gould. 1972. "Punctuated Equilibria: An Alternative to Phyletic Gradualism." *Paleobiology* 3 (2): 115–51.

Elliott, J. H. 1992. *The Old World and the New, 1492–1650*. New York: Cambridge University Press.

Erdrich, L. 2005. *The Painted Drum*. New York: HarperCollins.

Fabian, J. 1990. "Presence and Representation: The Other and Anthropological Writing." *Critical Inquiry* 16:753–72.

Fentress, J. C. 1967. "Observations on the Behavioral Development of a Hand-Reared Male Timber Wolf." *American Zoologist* 7:339–51.

Fogg, B. R. 2012. "The First Domestication: Examination of the Relationship between Indigenous *Homo sapiens* of America and Australia and *Canis lupus*." MA thesis, University of Kansas.

Fogg, B. R., N. Howe, and R. Pierotti. 2015. "Relationships between Indigenous American Peoples and Wolves, 1: Wolves as Teachers and Guides." *Journal of Ethnobiology* 35:262–85.

Foltz, R. 2006. *Animals in Islamic Tradition and Western Cultures*. Oxford: Oneworld.

Forsyth, J. 1992. *A History of the Peoples of Siberia: Russia's North Asian Colony, 1581–1990*. New York: Cambridge University Press.

Fouts, R., and S. T. Mills. 1997. *Next of Kin: What Chimpanzees Have Taught Me about Who We Are*. New York: William Morrow.

Fox, M. W. 1971. *Behaviour of Wolves, Dogs, and Related Canids*. New York: Harper and Row.

Francis, R. 2015. *Domesticated: Evolution in a Man-made World*. New York: Norton.

Franklin, J. 2009. *The Wolf in the Parlor: How the Dog Came to Share Your Brain*. New York: St. Martins Griffin.

Frantz, L. A. F., V. E. Mullin, M. Pionnier-Capitan, O. Labrasseur, M. Ollivier, et al. 2016. "Genomic and Archaeological Evidence Suggests a Dual Origin of Domestic Dogs." *Science* 352:1228–31.

Freedman, A. H., I. Gronau, R. M. Schweizer, D. Ortega-Del Vecchyo, E. Han, et al. 2014. "Genome Sequencing Highlights the Dynamic Early History of Dogs." *PLoS Genet* 10 (1): e1004016. doi:10.1371/journal.pgen.1004016.

Gade, G. 2002. *Wolves, Dogs, Hybrids and Plains Indians.* http://vorebuffalojump. org/pdf/Wolves,%20dogs,%20hybrids%20and%20Indians.pdf.

Garfield, V., and L. Forrest. 1961. *The Wolf and the Raven: Totem Poles of Southeast Alaska.* Seattle: University of Washington Press.

Garrigan, D., and M. E. Hammer. 2006. "Reconstructing Human Origins in the Genomic Era." *Nature Reviews Genetics* 7:669–80.

Germonpré, M., M. Lázničková-Galetová, and M. V. Sablin. 2012. "Palaeolithic Dog Skulls at the Gravettian Předmostí Site, the Czech Republic." *Journal of Archaeological Science* 39 (1): 184–202.

Germonpré, M., M. V. Sablin, V. Despre, M. Hofreiter, M. Lázničková-Galetová, R. E. Stevens, and M. Stiller. 2013. "Paleolithic Dogs and the Early Domestication of the Wolf: A Reply to the Comments of Crockford and Kuzmin." *Journal of Archaeological Science* 40:786–92.

Germonpré, M., M. V. Sablin, M. Lázničková-Galetová, V. Despre, R. E. Stevens, M. Stiller, and M. Hofreiter. 2015. "Palaeolithic Dogs and Pleistocene Wolves Revisited: A Reply to Morey." *Journal of Archaeological Science* 54:210–16.

Germonpré, M., M. V. Sablin, R. E. Stevens, R. E. M. Hedges, M. Hofreiter, et al. 2009. "Fossil Dogs and Wolves from Palaeolithic Sites in Belgium, the Ukraine and Russia: Osteometry, Ancient DNA and Stable Isotopes." *Journal of Archaeological Science* 36:473–90.

Gibbons, A. 2011. "Who Were the Denisovans?" *Science* 333:1084–87.

———. 2013. "How a Fickle Climate Made Us Human." *Science* 341:474–79.

———. 2015. "Humans May Have Reached Chile by 18,500 Years Ago." *Science* 350:898.

Gilbert, B. 1989. *God Gave Us This Country: Tekamthi and the First American Civil War.* New York: Anchor Books.

Goebel, T. 2004. "The Early Upper Paleolithic of Siberia." In *The Early Upper Paleolithic beyond Western Europe,* edited by P. J. Brantingham, S. L. Kuhn, and K. W. Kerry, 162–95. Berkeley: University of California Press.

Golden, P. B. 1997. "Wolves, Dogs and Qipcaq Religion." *Acta Orientalia Academiae Scientiarum Hungaricae. Tomus L* 1–3:87–97.

Goldizen, A. 1987. "Tamarins and Marmosets: Communal Care of Offspring." In *Primate Societies,* edited by B. Smuts, D. Cheney, R. Seyfarth, R. Wrangham, and T. Struhsaker, 34–43. Chicago: University of Chicago Press.

Good, T. P., J. Ellis, C. A. Annett, and R. Pierotti. 2000. "Bounded Hybrid Superiority: Effects of Mate Choice, Habitat Selection, and Diet in an Avian Hybrid Zone." *Evolution* 54:1774–83.

Gould, S. J. 2003. *The Hedgehog, the Fox, and the Magister's Pox.* New York: Harmony Books.

Gould, S. J., and R. C. Lewontin. 1979. "The Spandrels of San Marco and the Panglossian Paradigm: A Critique of the Adaptationist Programme." In "The Evolution of Adaptation by Natural Selection." Special issue, *Proceedings of the Royal Society of London, Series B, Biological Sciences* 205 (1161): 581–98.

Grant, P. R., and B. R. Grant. 1992. "Hybridization of Bird Species." *Science* 256:193–97.

———. 1997a. "Genetics and the Origin of Bird Species." *Proceedings of the National Academy of Science* 94:7768–75.

———. 1997b. "Hybridization, Sexual Imprinting, and Mate Choice." *American Naturalist* 149:1–28.

Gray, J. 2002. *Straw Dogs: Thoughts on Humans and Other Animals.* New York: Farrar, Straus, and Giroux.

Grimaud, H. 2003. *Wild Harmonies: A Life of Music and Wolves.* Translated by Ellen Hinsey. New York: Riverhead Books.

Grinnell, G. B. 1926. *By Cheyenne Campfires.* Lincoln: University of Nebraska Press.

———. 1972. *Blackfoot Lodge Tales.* Williamstown, MA: Corner House. Originally published in 1892.

Guiler, E. 1985. *Thylacine: The Tragedy of the Tasmanian Tiger.* Melbourne: Oxford University Press.

Haag, W. G. 1956. "Aboriginal Dog Remains from Yellowstone National Park." Yellowstone National Park Archives, Yellowstone National Park, WY.

Haber, G., and M. Holleman. 2013. *Among Wolves: Gordon Haber's Insights into Alaska's Most Misunderstood Animal.* Fairbanks: University of Alaska Press.

Hall, R. L., and H. S. Sharp, eds. 1978. *Wolf and Man: Evolution in Parallel.* New York: Academic Press.

Hammer, M., et al. 2011. "Genetic Evidence for Archaic Admixture in Africa." *Proceedings of the National Academy of Sciences, USA* 108:15123–28.

Hampton, B. 1997. *The Great American Wolf.* New York: Henry Holt.

Hand, J. L. 1986. "Resolution of Social Conflicts: Dominance, Egalitarianism, Spheres of Dominance, and Game Theory." *Quarterly Review of Biology* 61:201–20.

Hare, B., M. Brown, C. Williamson, and M. Tomasello. 2002. "The Domestication of Social Cognition in Dogs." *Science* 298:1634–36.

Hare, B., and V. Woods. 2013. *The Genius of Dogs: How Dogs Are Smarter Than You Think.* New York: Dutton Books.

Harney, C. 1995. *The Way It Is: One Water—One Air—One Mother Earth.* Nevada City, CA: Blue Dolphin.

Harris, E. E. 2015. *Ancestors in Our Genome: The New Science of Human Evolution.* New York: Oxford University Press.

Hassrick, Royal B. 1964. *The Sioux: Life and Customs of a Warrior Society.* Norman: University of Oklahoma Press.

Head, J. J., P. M. Barrett, and E. J. Rayfield. 2009. "Neurocranial Osteology and Systematic Relationships of *Varanus (Megalania) prisca* Owen, 1859 (Squamata: Varanidae)." *Zoological Journal of the Linnean Society* 155:445–57.

Hearne, S. 1958. *A Journey from the Prince of Wales's Fort in Hudson Bay to the Northern Ocean, 1769–1772*. Toronto: Macmillan.

Heinrich, B. 1999. *The Mind of the Raven: Investigations and Adventures with Wolf-Birds*. New York: HarperCollins.

Hemmer, Helmut. 1990. *Domestication: The Decline of Environmental Appreciation*. Cambridge: Cambridge University Press.

Hernandez, N. 2013. "The Other Becomes the Self: Reciprocity and Reflection in Land-Animal-Human Ecologies." Paper presented at the American Society of Literature and the Environment Conference, Lawrence, KS, May 28–June 1.

———. 2014. " 'Wolf Man' and Wolf Knowledge in Native American Hunting Traditions." Paper presented at the Society of Ethnobiology Meeting, Cherokee, NC.

Herzog, W., and D. Vasyukov. 2010. *The Happy People: A Year in the Taiga*. Potsdam, Germany: Studio Babelsberg.

Hess, D. J. 1995. *Science and Technology in a Multicultural World: The Cultural Politics of Facts and Artifacts*. New York: Columbia University Press.

Hinde, R. A. 1976. "Interactions, Relationships and Social Structure." *Man* 11:1–17.

Hobbes, T. 1985. *Leviathan; or, The Matter, Forme, & Power of a Common-wealth, Ecclesiasticall and Civill*. Edited by C. B. Macpherson. Harmondsworth, UK: Penguin. Originally published in 1651.

Hoffman, W., D. Heinemann, and J. A. Wiens. 1981. "The Ecology of Seabird Feeding Flocks in Alaska." *Auk* 98:437–56.

Honacki, J. H., K. E. Kinman, and J. W. Koeppl, eds. 1982. *Mammal Species of the World: A Taxonomic and Geographic Reference*. Lawrence, KS: Allen.

Hope, J. 1994. "Wolves and Wolf Hybrids as Pets Are Big Business—but a Bad Idea." *Smithsonian* 25 (3): 34–45.

Hunn, E. S. 2013. " 'Dog' as Life Form." In *Explorations in Ethnobiology: The Legacy of Amadeo Rea*, edited by M. Quinlan and D. Lepofsky, 141–53. Society of Ethnobiology Contributions in Ethnobiology.

Hyde, G. E. 1968. *A Life of George Bent, Written from His Letters*. Edited by S. Lottinville. Norman: University of Oklahoma Press, OK.

Hyde, L. 1998. *Trickster Makes the World: Mischief, Myth, and Art*. New York: North Point.

Ioannesyan, A. R. 1990. *Materials on the Cynology of Hunting Dogs* [in Russian]. Moscow: Rosokhotrybolovsoyuz.

Itard, Jean-Marc-Gaspard. 1962. *The Wild Boy of Aveyron*. New York: Meredith.

Jans, N. 2015. *A Wolf Called Romeo*. New York: Mariner Books.

Janson, C. H., and M. Goldsmith. 1995. "Predicting Group Size in Primates: Foraging Costs and Predation Risks." *Behavioral Ecology* 6:326–36.

Kappeler, P. M., and E. W. Heymann. 1996. "Nonconvergence in the Evolution of Primate Life History and Socio-ecology." *Biological Journal of the Linnaean Society* 59:297–326.

Kappeler, P. M., and C. P. van Schaik. 2002. "Evolution of Primate Social Systems." *International Journal of Primatology* 23 (4): 707–40.

Kean, S. 2012. *The Violinist's Thumb and Other Tales of Love, War, Genius, as Written by Our Genetic Code.* New York: Little Brown.

Kirschner, M. W., and J. C. Gerhart. 2005. *The Plausibility of Life: Resolving Darwin's Dilemma.* New Haven: Yale University Press.

Kluckhorn, C. 1944. *Navaho Witchcraft.* Boston: Beacon.

Knight, J. 2012. "Chippewa Tribes Oppose State Wolf Hunt." *Eau Claire Leader Telegram,* August 15. http://www.leadertelegram.com/news/front_page/article_188174ec-e69a-11e1-a079-001a4bcf887a.html?mode=story.

Krings, M., et al. 1997. "Neanderthal DNA Sequences and the Origins of Modern Humans." *Cell* 90:19–30.

Kuhn, T. 1996. *The Structure of Scientific Revolutions.* 3rd ed. Chicago: University of Chicago Press.

Kurz, F. 1937. "Journal of Rudolph Friedrich Kurz." *Bureau of American Ethnology Bulletin,* no. 115.

Lane, H. 1976. *The Wild Boy of Aveyron.* Cambridge, MA: Harvard University Press.

Lawson, J. 1967. *A New Voyage to Carolina: Containing the Exact Description and Natural History of That Country; Together with the Present State Thereof. And a Journal of a Thousand Miles, Travel'd thro; Several Nations of Indians. Giving a Particular Account of Their Customs, Manners, &c.* Edited by H. T. Lefler. Chapel Hill: University of North Carolina Press.

Leach, H. M. 2003. "Human Domestication Reconsidered." *Current Anthropology* 44 (3): 349–68.

Leonard, J. A., R. K. Wayne, J. Wheeler, R. Valadez, S. Guillen, and C. Vila. 2002. "Ancient DNA Evidence for Old World Origin of New World Dogs." *Science* 298:1613–16.

Lévi-Strauss, C. 1967. *Structural Anthropology.* New York: Doubleday.

Lewis, R. 2013. "Wisconsin Wolf Hunt Begins amid Warnings from Conservationists, Tribes." *Al Jazeera,* October 16.

Lewontin, R. 2001. *The Triple Helix: Gene, Organism, and Environment.* Cambridge, MA: Harvard University Press.

Lienhard, J. H. 1998–99. "Dogs." *Engines of Our Ingenuity,* no. 1431. http://www.uh.edu/engines/epi1431.htm.

Lincoln, W. B. 1994. *The Conquest of a Continent: Siberia and the Russians.* New York: Random House.

Linnaeus, C. 1792. *The Animal Kingdom; or, Zoological System of the Celebrated Sir Charles Linnaeus. Class I. Mammalia and Class II. Birds. Being a Translation of That Part of the "Systema Naturae," as Lately Published with Great Improvements by Professor Gmelin, Together with Numerous Additions from More Recent Zoological Writers and Illustrated with Copperplates.* Translated and edited by R. Kerr. London: J. Murray.

Loendorf, L. L., and N. M. Stone. 2006. *The Sheep Eater Indians of Yellowstone.* Salt Lake City: University of Utah Press.

Lopez, B. H. 1978. *Of Wolves and Men.* New York: Charles Scribner.

Lorenzen, E. D., D. Nogues-Bravo, et al. 2011. "Species-Specific Responses of Late Quaternary Megafauna to Climate and Humans." *Nature* 479:359–63.

Losey, R. J., S. Garvie-Lok, J. A. Leonard, M. A. Katzenberg, M. Germonpré, et al. 2013. "Burying Dogs in Ancient Cis-Baikal, Siberia: Temporal Trends and Relationships with Human Diet and Subsistence Practices." *PLoS ONE* 8 (5): e63740. doi:10.1371/journal.pone.0063740.

Lowie, R. H. 1909. "The Northern Shoshone." *Anthropological Papers of the American Museum of Natural History* 2:165–306.

Lumholtz, C. 1889. *Among Cannibals: An Account of Four Years' Travels in Australia and of Camp Life with the Aborigines of Queensland.* New York: Charles Scribner's Sons.

Major, P. F. 1978. "Predator-Prey Interactions in Two Schooling Fishes, *Caranx ignobilis* and *Stolephorus purpureus.*" *Animal Behaviour* 26:760–77.

Malone, N., A. Fuentes, and F. J. White. 2012. "Variation in the Social Systems of Extant Hominoids: Comparative Insight into the Social Behavior of Early Hominins." *International Journal of Primatology* 25:97–164. doi:10.1007/s10764-012-9617-0.

Mann, C. C. 2002. "1491." *Atlantic Monthly,* March, 41–53.

———. 2005. *1491: New Revelations of America before Columbus.* New York: Knopf.

Margulis, L. 1998. *Symbiotic Planet: A New View of Evolution.* New York: Basic Books.

Marshall, Joseph III. 1995. *On Behalf of the Wolf and the First Peoples.* Santa Fe: Red Crane.

———. 2001. *The Lakota Way: Stories and Lessons for Living.* New York: Viking Compass.

———. 2005. *Walking with Grandfather: The Wisdom of Lakota Elders.* Boulder, CO: Sounds True.

Marshall Thomas, E. 1993. *The Hidden Life of Dogs.* Boston: Houghton Mifflin.

———. 1994. *The Tribe of Tiger: Cats and Their Culture.* New York: Simon and Schuster.

———. 2000. *The Social Lives of Dogs: The Grace of Canine Company.* New York: Simon and Schuster.

———. 2006. *The Old Way: A Story of the First People.* New York: Farrar Straus Giroux.

———. 2013. *A Million Years with You: A Memoir of Life Observed.* New York: Houghton Mifflin Harcourt.

Martin, C. L. 1999. *The Way of the Human Being.* New Haven: Yale University Press.

Mason, P. H., and R. V. Short. 2011. "Neanderthal-Human Hybrids." *Hypothesis* 9 (1): e1.

Maximilian, A. P. 1906. *Travels in the Interior of North America.* Edited by R. G. Thwaites. Cleveland: A. C. Clark.

McCaig, D. 1991. *Eminent Dogs, Dangerous Men: Searching through Scotland for a Border Collie.* New York: Edward Burlingame Books.

McClintock, W. 1910. *The Old North Trail.* London: Macmillan.

McFall-Ngai, M., M. G. Hadeld, T. C. G. Bosch, H. V. Carey, T. Domazet-Loso, A. E. Douglas, N. Dubilier, G. Eberl, T. Fukami, S. F. Gilbert, U. Hentschel,

N. King, S. Kjelleberg, A. H. Knoll, N. Kremer, S. K. Mazmanian, J. L. Metcalf, K. Nealson, N. E. Pierce, J. F. Rawls, A. Reid, E. G. Ruby, M. Rumpho, J. G. Sanders, D. Tautz, and J. J. Wernegreen. 2013. "Animals in a Bacterial World: A New Imperative for the Life Sciences." *Proceedings of the National Academy of Sciences* 110:3229–36.

McHenry, H. M. 1992. "Body Size and Proportions in Early Hominids." *American Journal of Physical Anthropology* 87:407–31.

———. 1996. "Sexual Dimorphism in Fossil Hominids and Its Socioecological Implications." In *The Archaeology of Human Ancestry Power, Sex, and Tradition*, edited by J. Steele and S. Shennan, 91–109. London: Routledge.

McHenry, H. M., and K. Coffing. 2000. "Australopithecus to Homo: Transformations in Body and Mind." *Annual Review of Anthropology* 29:125–46.

McIntyre, Rick. 1995. *War against the Wolf: America's Campaign to Exterminate the Wolf.* Stillwater, MN: Voyageur.

McLeod, P. J. 1990. "Infanticide by Female Wolves." *Canadian Journal of Zoology* 68:402–4.

Mech, L. D. 1970. *The Wolf: The Ecology and Behavior of an Endangered Species.* New York: Natural History.

———. 1995. "A Ten-Year History of the Demography and Productivity of an Arctic Wolf Pack." *Arctic* 48:329–32.

———. 1999. "Alpha Status, Dominance, and Division of Labor in Wolf Packs." *Canadian Journal of Zoology* 77:1196–1203.

Mech, L. D., L. G. Adams. T. J. Meier, J. W. Burch, and B. W. Dale. 1998. *The Wolves of Denali.* Minneapolis: University of Minnesota Press.

Medin, D. L., and M. Bang. 2014. *Who's Asking? Native Science, Western Science, and Science Education.* Cambridge, MA: MIT Press.

Meggitt, M. J. 1965. "Association between Australian Aborigines and Dingoes." In *Man, Culture, and Animals: The Role of Animals in Human Ecological Adjustments*, edited by A. Leeds and A. P. Vayda, 7–26. Washington, DC: [American Association for the Advancement of Science].

Merial Limited. 2008. "The Benefits of IMRAB." http://imrab.us.merial.com /imrab/index.shtml.

Miklósi, Á. 2007. *Dog: Behavior, Evolution and Cognition.* New York: Oxford University Press.

Miklósi, Á., E. Kubinyi, J. Topál, M. Gácsi, Z. Virányi, and V. Csányi. 2003. "A Simple Reason for a Big Difference: Wolves Do Not Look Back at Humans but Dogs Do." *Current Biology* 13:763–66.

Miklósi, Á., R. Polgárdi, J. Topál, and V. Csányi. 1998a. "An Experimental Analysis of 'Showing' Behaviour in the Dog." *Animal Cognition* 3:159–66.

———. 1998b. "Use of Experimenter-Given Cues in Dogs." *Animal Cognition* 1:113–21.

Millan, C., and M. J. Peltier. 2006. *Cesar's Way: The Natural, Everyday Guide to Understanding and Correcting Common Dog Problems.* New York: Harmony Books.

Miller, W. 1972. *Newe Natekwinappeh: Shoshoni Stories and Dictionary.* Anthropological Papers 94. University of Utah Anthropology Department.

Minta, S. C., K. A. Minta, and D. F. Lott. 1992. "Hunting Associations between Badgers and Coyotes." *Journal of Mammalogy* 73:814–20.

Mitani, J. 1990. "Experimental Field Studies of Asian Ape Social Systems." *International Journal of Primatology* 11 (2): 103–26.

Mitani, J. C., J. Gros-Louis, and J. H. Manson. 1996 "Number of Males in Primate Groups: Comparative Tests of Competing Hypotheses." *American Journal of Primatology* 38:315–32.

Moehlman, P. D. 1989. "Intraspecific Variation in Canid Social Systems." In *Carnivore Behavior, Ecology, and Evolution*, edited by J. Gittleman, 143–63. Ithaca: Cornell University Press.

Moore, R. E. 2015. *The Tonkawa Indians*. http://www.texasindians.com/tonk.htm.

Morey, D. 1986. "Studies on Amerindian Dogs: Taxonomic Analysis of Canid Crania from the Northern Plains." *Journal of Archaeological Science* 13:119–45.

———. 1990. "Cranial Allometry and the Evolution of the Domestic Dog." PhD diss., University of Tennessee, Knoxville.

———. 1994. "The Early Evolution of the Domestic Dog." *American Scientist* 82:336–47.

———. 2010. *Dogs: Domestication and the Development of a Social Bond*. New York: Cambridge University Press.

———. 2014. "In Search of Paleolithic Dogs: A Quest with Mixed Results." *Journal of Archaeological Science* 52:300–307.

Morey, D., and R. Jeger. 2015. "Paleolithic Dogs: Why Sustained Domestication Then?" *Journal of Archaeological Science: Reports* 3:420–28.

Morris, D. 2001. *Dogs: The Ultimate Dictionary of Over 1000 Dog Breeds*. North Pomfret, VT: Trafalgar Square.

Morse, E. S. 1945. *Japan Day by Day: 1877, 1878–1879, 1882–1883*. Vol. 2. Boston: Houghton Mifflin.

Mourning Dove and H. D. Guie. 1990. *Coyote Stories*. Lincoln: University of Nebraska Press. Originally published in 1933.

Muir, J. 1916. *The Writings of John Muir*. Vol. 3. Boston: Houghton Mifflin.

Munday, P. 2013. "Thinking through Ravens: Human Hunters, Wolf-Birds, and Embodied Communication." In *Perspectives on Human-Animal Communication: Internatural Communication*, edited by E. Plec, 207–25. New York: Routledge.

National Geographic. 2013. "Solo: The Wild Dog's Tale." *Inside Wild*, March 29. tvblogs.nationalgeographic.com/2013/03/29/solo-the-wild-dogs-tale/.

Nelson, R. 1983. *Make Prayers to the Raven*. Chicago: University of Chicago.

Nickerson, P. 1997. Statement presented to Michigan House of Representatives, Lansing.

Nisbet, J., and C. Nisbet. 2010. "Mourning Dove (Christine Quintasket) (ca. 1884–1936)." Historylink.org. http://www.historylink.org/File/9512.

Niskanen, A. K., E. Hagstrom, H. Lohi, M. Ruokonen, R. Esparza-Salas, J. Aspi, and P. Savolaeinen. 2013. "MNC Variability Supports Dog Domestication from a Large Number of Wolves: High Diversity in Asia." *Heredity* 110:80–85.

Nowack, E. C., and M. Melkonian. 2010. "Endosymbiotic Associations within Protists." *Philosophical Transactions of the Royal Society B: Biological Sciences* 365 (1541): 699–712.

Nowak, R. 1979. *North American Quaternary Canis.* Lawrence: University of Kansas Monographs.

O'Brien, S. J., and E. Mayr. 1991. "Bureaucratic Mischief: Recognizing Endangered Species and Subspecies." *Science* 251:1187–88.

O'Connor, T., and H. Yu Wong. 2012. "Emergent Properties." In *The Stanford Encyclopedia of Philosophy,* edited by E. N. Zalta. http://plato.stanford.edu /archives/spr2012/entries/properties-emergent/.

Ohlson, D. L. 2005. "Tribal Sovereignty and the Endangered Species Act: Recovering the Idaho Wolves." MA thesis, San Jose State University.

Olsen, S. 1985. *Origins of the Domestic Dog: The Fossil Record.* Tucson: University of Arizona Press.

Ovodov, N. D., S. J. Crockford, Y. V. Kuzmin, T. F. G. Higham, G. W. L. Hodgins, and J. van der Plicht. 2011. "A 33,000 Year-Old Incipient Dog from the Altai Mountains of Siberia: Evidence of the Earliest Domestication Disrupted by the Last Glacial Maximum." *PLoS ONE* 6 (7): e22821. doi:10.1371/journal. pone.0022821.

Owen, D. 2003. *Tasmanian Tiger: The Tragic Tale of How the World Lost Its Most Mysterious Predator.* Baltimore: Johns Hopkins University Press.

Owen, R. 1859a. "*Megalainia priscus.*" *Philosophical Transactions of the Royal Society of London* 149:43–48.

———. 1859b. "On the Fossil Mammals of Australia, Part II: Description of a Mutilated Skull of the Large Marsupial Carnivore (*Thylacoleo carnifex* Owen), from a Calcareous Conglomerate Stratum, Eighty Miles S.W. of Melbourne, Victoria." *Philosophical Transactions of the Royal Society* 149:309–22.

Packer, C., and L. Ruttan. 1988. "The Evolution of Cooperative Hunting." *American Naturalist* 132:159–98. doi:10.1086/284844.

Papanikolas, Z. 1995. *Trickster in the Land of Dreams.* Lincoln: Bison Books, University of Nebraska Press.

Pastoureau, M. 2011. *The Bear: History of a Fallen King.* Cambridge, MA: Belknap Press, Harvard University Press.

Pavlik, S. 1999. "San Carlos and White Mountain Apache Attitudes toward the Reintroduction of the Mexican Wolf to Its Historic Range in the American Southwest." *Wicazo Sa Review* 14:129–45.

Pennisi, E. 2004. "Ice Ages May Explain Ancient Bison's Boom-Bust History." *Science* 306:1454.

———. 2006. "Social Animals Prove Their Smarts." *Science* 312:1734–38.

People of Michigan. 2000. Wolf-Dog Cross Act: Act 246 of 2000. http://www. legislature.mi.gov/%28S%28yxp3ts55y131viq313mpok45%29%29/documents/ mcl/pdf/mcl-Act–246-of–2000.pdf.

Peterson, R. O. 1995. *The Wolves of Isle Royale: A Broken Silence.* Minocqua, WI: Willow Creek.

Petto, A. J., and L. R. Godfrey. 2007. *Scientists Confront Intelligent Design and Creationism*. New York: Norton.

Pierotti, R. 1981. "Male and Female Parental Roles in the Western Gull under Different Environmental Conditions." *Auk* 98:532–49.

———. 1982. "Habitat Selection and Its Effect on Reproductive Output in the Herring Gull in Newfoundland." *Ecology* 63:854–68.

———. 1988a. "Associations between Marine Birds and Marine Mammals in the Northwest Atlantic." In *Seabirds and Other Marine Vertebrates: Commensalism, Competition, and Predation*, edited by J. Burger, 31–58. New York: Columbia University Press.

———. 1988b. "Interactions between Gulls and Otariid Pinnipeds: Competition, Commensalism, and Cooperation." In *Seabirds and Other Marine Vertebrates: Commensalism, Competition, and Predation*, edited by J. Burger, 213–39. New York: Columbia University Press.

———. 2004. "Animal Disease as an Environmental Factor." In *Encyclopedia of World Environmental History*, edited by S. Krech and C. Merchant. New York: Berkshire.

———. 2006. "The Role of Animal Disease in History." In *Encyclopedia of World History*. New York: Berkshire.

———. 2010. "Sustainability of Natural Populations: Lessons from Indigenous Knowledge." *Human Dimensions of Wildlife* 15:274–87.

———. 2011a. *Indigenous Knowledge, Ecology, and Evolutionary Biology*. New York: Routledge.

———. 2011b. "The World According to Is'a: Combining Empiricism and Spiritual Understanding in Indigenous Ways of Knowing." In *Ethnobiology*, edited by E. N. Anderson, D. M. Pearsall, E. S. Hunn, and N. J. Turner, 65–82. Hoboken, NJ: Wiley-Blackwell.

———. 2012a. "All Dogs Are Wolves." *Wolfdog* 2:8–11.

———. 2012b. "The Process of Domestication: Why Domestic Forms Are Not Species." *Wolfdog* 2:22–27.

———. 2014. "A Tapestry, Not a Tree." *Science Online* comments, February 15. http://comments.sciencemag.org/content/10.1126/science.1243650 #comments.

Pierotti, R., and C. A. Annett. 1993. "Hybridization and Male Parental Care in Birds." *Condor* 95:670–79.

———. 1994. "Patterns of Aggression in Gulls: Asymmetries and Tactics in Different Roles." *Condor* 96:590–99.

Pierotti, R., C. A. Annett, and J. L. Hand. 1996. "Male and Female Perceptions of Pair-Bond Dynamics: Monogamy in the Western Gull." In *Feminism and Evolutionary Biology*, edited by P. A. Gowaty, 261–75. New York: Chapman and Hall.

Pierotti, R., and D. Wildcat. 1997. "The Science of Ecology and Native American Traditions." *Winds of Change* 12 (4): 94–98.

———. 2000. "Traditional Ecological Knowledge: The Third Alternative." *Ecological Applications* 10:1333–40.

Plyusnina, I., et al. 1991. "An Analysis of Fear and Aggression during Early Development of Behavior in Silver Foxes (*Vulpes vulpes*)." *Applied Animal Behavior Science* 32:253–68.

Porter, J. M., and S. G. Sealy. 1981. "Dynamics of Seabird Multispecies Feeding Flocks: Chronology of Flocking in Barkley Sound, British Columbia, in 1979." *Colonial Waterbirds* 4:104–13.

———. 1982. "Dynamics of Seabird Multispecies Feeding Flocks: Age-Related Feeding Behaviour." *Behaviour* 81:91–109.

Powell, P. J. 1979. *Sweet Medicine: Continuing Role of the Sacred Arrows, the Sun Dance, and the Sacred Buffalo Hat in Northern Cheyenne History*. Civilization of the American Indian Series. Norman: University of Oklahoma Press.

Prindle, D. F. 2009. *Stephen J. Gould and the Politics of Evolution*. Amherst, NY: Prometheus.

Prothero, D. R. 2007. *Evolution: What the Fossils Say and Why It Matters*. New York: Columbia University Press.

Pryor, K., J. Lindbergh, S. Lindbergh, and R. Milano. 1990. "A Dolphin-Human Fishing Cooperative in Brazil." *Marine Mammal Science* 6 (1): 77–82.

Quinney, P. S., and M. Collard. 1995. "Sexual Dimorphism in the Mandible of *Homo neanderthalensis* and *Homo sapiens*: Morphological Patterns and Behavioural Implications." *Archaeological Sciences* 1995:420–25.

Radin, P. 1972. *The Trickster: A Study in American Indian Mythology*. New York: Schocken Books.

Ramsey, J. 1977. *Coyote Was Going There: Indian Literature of the Oregon Country*. Seattle: University of Washington Press.

———. 1999. *Reading the Fire: The Traditional Indian Literatures of America*. Seattle: University of Washington Press.

Ray, D. 1967. *Eskimo Masks: Art and Ceremony*. Seattle: University of Washington Press.

Richard, A. F. 1978. *Behavioral Variation: Case Study of a Malagasv Lemur*. Lewisburg, PA: Bucknell University Press.

Ricketts, M. 1966. "The North American Indian Trickster." *History of Religions* 5:323–50.

Rindos, D. 1984. *The Origins of Agriculture: An Evolutionary Perspective*. Orlando: Academic Press.

Ritvo, H. 2010. *Nobel Cows and Hybrid Zebras: Essays on Animals and History*. Charlottesville: University of Virginia Press.

Roff, D. 1992. *The Evolution of Life Histories: Theory and Analysis*. London: Chapman and Hall.

Rose, D. Bird. 1996. *Nourishing Terrains: Australian Aboriginal Views of Landscape and Wilderness*. Canberra: Australian Heritage Commission.

———. 2000. *Dingo Makes Us Human: Life and Land in an Aboriginal Australian Culture*. Cambridge: Cambridge University Press.

———. 2011. *Wild Dog Dreaming: Love and Extinction*. Charlottesville: University of Virginia Press.

Rothstein, S. I., and R. Pierotti. 1988. "Distinctions among Reciprocal Altruism and Kin Selection, and a Model for the Initial Evolution of Helping Behavior." *Ethology and Sociobiology* 9:189–210.

Roughsey, R. 1973. *The Giant Devil Dingo.* http://www.austlit.edu.au/austlit/page /C317845.

Russell, E. 2011. *Evolutionary History: Uniting History and Biology to Understand Life on Earth.* New York: Cambridge University Press.

Russell, O. 1914. *Journal of a Trapper; or, Nine Years in the Rocky Mountains, 1834–1843.* Edited by L. A. York. Boise, ID: Syms York.

Sabaneev, L. P. 1993. *Hunting Dogs: Sight Hounds and Scent Hounds* [in Russian]. Moscow: Terra.

Sablin, M. V., and G. A. Khlopachev. 2002. "The Earliest Ice Age Primitive Dogs Are Found at the Russian Upper Paleolithic Site Eliseevichi." *Current Anthropology* 43:795–99.

Sacks, J. J., L. Sinclair, J. Gilchrist, G. C. Golab, and R. Lockwood 2000. "Breeds of Dog Involved in Fatal Human Attacks in the United States between 1979 and 1998." *Journal of the American Coates Medical Association* 217:836–43.

Safina, C. 2015. *Beyond Words: What Animals Think and Feel.* New York: Henry Holt.

Sale, K. 1991. *The Conquest of Paradise: Christopher Columbus and the Columbian Legacy.* New York: Knopf.

Samar, A. P. 2010. "Traditional Dog-Breeding of the Nanai People." *Journal of the International Society for the Preservation of Primitive and Aboriginal Dogs* 34:19–51.

Sankararaman, S., et al. 2012. "The Date of Interbreeding between Neanderthals and Modern Humans." *PLoS Genetics* 8: e1002947.

Savolainen, P., Y. P. Zhang, J. Luo, J. Lundeberg, and T. Leitner. 2002. "Genetic Evidence for an East Asian Origin of Domestic Dogs." *Science* 298:1610–13.

Schleidt, W. M. 1998. "Is Humaneness Canine?" *Human Ethology Bulletin* 13 (4): 3–4.

Schleidt, W. M., and M. D. Shalter. 2003. "Co-evolution of Humans and Canids: An Alternative View of Dog Domestication: Homo Homini Lupus?" *Evolution and Cognition* 9 (1): 57–72.

Schlesier, K. H. 1987. *The Wolves of Heaven: Cheyenne Shamanism, Ceremonies, and Prehistoric Origins.* Civilization of the American Indian 183. Norman: University of Oklahoma Press.

Schullery, P., and B. Babbitt. 2003. *The Yellowstone Wolf: A Guide and Sourcebook.* Norman: University of Oklahoma Press.

Schwartz, M. 1997. *A History of Dogs in the Early Americas.* New Haven: Yale University Press.

Scott, J. P. 1968. "Evolution and Domestication of the Dog." *Evolutionary Biology* 2:243–75.

Serpell, J., ed. 1995. *The Domestic Dog.* Cambridge: Cambridge University Press.

Shapiro, B., A. J. Drummond, et al. 2004. "Rise and Fall of the Beringian Steppe Bison." *Science* 306:1561–64.

Shavit, A., and J. R. Griesemer. 2011. "Mind the Gaps: Why Are Niche Construction Processes So Rarely Used?" In *Lamarckian Transformations*, edited by S. Gisis and E. Jablonka, 307–17. Cambridge, MA: MIT Press.

Shimkin, D. 1939. "Interactions of Culture, Needs, and Personalities among the Wind River Shoshone." PhD diss., UCLA.

———. 1986. "Eastern Shoshone." In *Handbook of the North American Indians,* vol. 11, *Great Basin,* edited by W. d'Azevedo, 308–35. Washington, DC: Smithsonian Institution Press.

Shipman, P. 2011. *The Animal Connection.* New York: Norton.

———. 2014. "How Do You Kill 86 Mammoths? Taphonomic Investigations of Mammoth Megasites." *Quaternary International* 359–60:1–9.

———. 2015. *The Invaders: How Humans and Their Dogs Drove Neanderthals to Extinction.* Cambridge, MA: Harvard University Press.

Shouse, B. 2003. "Ecology: Conflict over Cooperation." *Science* 299:644–46.

Shreeve, J. 2013. "The Case of the Missing Ancestor: DNA from a Cave in Russia Adds a Mysterious New Member to the Human Family." *National Geographic,* July. http://ngm.nationalgeographic.com/2013/07/125-missing-human-ancestor/shreeve-text.

Simeon, A. 1977. *The She-Wolf of Tsla-a-wat: Indian Stories for the Young.* Vancouver: J. J. Douglas.

Simpson, L. 2008. "Looking after Gdoo-naaganinaa: Precolonial Nishnaabeg Diplomatic and Treaty Relationships." *Wicaso Sa Review* (Fall): 29–42.

Singh, J. A. L., and R. M. Zingg. 1966. *Wolf-Children and Feral Man.* Hamden, CT: Archon Books. Originally published in 1942.

Skabelund, A. 2004. "Loyalty and Civilization: A Canine History of Japan." PhD diss., Columbia University.

Skoglund, P., E. Ersmark, E. Palkopoulou, and L. Dalén. 2015. "Ancient Wolf Genome Reveals an Early Divergence of Domestic Dog Ancestors and Admixture into High-Latitude Breeds." *Current Biology* 25:1–5.

Skoglund, P., A. Götherström, and M. Jakobsson. 2011. "Estimation of Population Divergence Times from Non-overlapping Genomic Sequences: Examples from Dogs and Wolves." *Molecular Biology and Evolution* 28 (4): 1505–17.

Smith, A. M., and A. C. Hayes. 1993. *Shoshone Tales.* Salt Lake City: University of Utah Press.

Smith, G. K. 1978. *Slave to a Pack of Wolves.* Chicago: Adams.

Spady, T. C., and E. A. Ostrander. 2008. "Canine Behavioral Genetics: Pointing out the Phenotypes and Herding Up the Genes." *American Journal of Human Genetics* 82:10–18.

Spotte, S. 2012. *Societies of Wolves and Free-ranging Dogs.* New York: Cambridge University Press.

Spottiswoode, C. N., K. S. Begg, and C. M. Begg. 2016. "Reciprocal Signaling in Honeyguide-Human Mutualism." *Science* 353 (6297): 387–89. doi:10.1126/science.aaf4885.

State v. Choate. 1998. Missouri Court of Appeals. http://law.justia.com/cases/missouri/court-of-appeals/1998/wd54407-2.html.

Stedman, H. H., B. W. Kozyak, A. Nelson, D. M. Thesier, L. T. Su, D. W. Low, C. R. Bridges, J. B. Shrager, N. M. Purvis, and M. A. Mitchell. 2004. "Myosin

Gene Mutation Correlates with Anatomical Changes in the Human Lineage." *Nature* 428:415–18.

Steenbeek, R., and C. P. van Schaik. 2001. "Competition and Group Size in Thomas's Langur (*Presbytis thomasi*): The Folivore Paradox Revisited." *Behavioral Ecology and Sociobiology* 49:100–110.

Stephenson, R. O. 1982. "Nunamiut Eskimos, Wildlife Biologists, and Wolves." In *Wolves of the World,* edited by F. H. Harrington and P. C. Pacquet, 434–39. Park Ridge, NJ: Noyes.

Strehlow, T. G. H. 1971. *Songs of Central Australia.* Sydney: Angus and Robertson.

Summers, M. 1966. *The Werewolf.* New Hyde Park, NY: University Books.

Swedell, L. 2012. "Primate Sociality and Social Systems." *Nature Education Knowledge* 3 (10): 84.

Tauber, A. I. 2009. *Science and the Quest for Meaning.* Waco, TX: Baylor University Press.

Terhune, A. P. 1935. *Real Tales of Real Dogs.* Akron, OH: Saalfield.

Terrill, C. 2011. *Part Wild: One Woman's Journey with a Creature Caught between the Worlds of Wolves and Dogs.* New York: Scribner.

Thalmann, O., et al. 2013. "Complete Mitochondrial Genomes of Ancient Canids Suggests a European Origin of Domestic Dogs." *Science* 342:871–74.

Thomas, D. H. 2000. *Skull Wars: Kennewick Man, Archaeology, and the Battle for Native American Identity.* New York: Basic Books.

Thomson, D. F. 1949. "Arnhem Land: Explorations among an Unknown People, Part III: On Foot across Arnhem Land." *Geographical Journal* 114:53–67.

Thurston, M. E. 1996. *The Lost History of the Canine Race.* Kansas City: Andrews McMeel.

Trivers, R. L. 1972. "Parental Investment and Sexual Selection." In *Sexual Selection and the Descent of Man, 1871–1971,* edited by B. Campbell, 136–79. Chicago: Aldine.

Trut, L. 1991. "Intracranial Allometry and Morphological Changes in Silver Foxes (*Vulpes vulpes*) under Domestication." *Genetika* 27:1605–11.

———. 1999. "Early Canid Domestication: The Farm-Fox Experiment." *American Scientist* 87:160–69.

———. 2001. "Experimental Studies of Early Canid Domestication." In *The Genetics of the Dog,* edited by A. Ruvinsky and J. Sampson, 15–42. New York: CABI.

Trut, L., et al. 2000. "Inter-hemispheric Biochemical Differences in Brains of Silver Foxes Selected for Behavior, and the Problem of Directional Asymmetry." *Genetika* 36:940–46.

Tsuda, K., Y. Kikkawa, H. Yonekawa, and Y. Tanabe. 1997. "Extensive Interbreeding Occurred among Multiple Matriarchal Ancestors during the Domestication of Dogs: Evidence from Inter- and Intraspecies Polymorphisms in the D-loop Region of Mitochondrial DNA between Dogs and Wolves." *Genes & Genetic Systems* 72:229–38.

Ude, W. 1981. *Becoming Coyote.* Amherst, MA: Lynx House.

Udell, M. A. R., N. R. Dorey, and C. D. L. Wynne. 2008. "Wolves Outperform Dogs in Following Human Social Cues." *Animal Behaviour* 76:1767–73.

Vander, J. 1997. *Shoshone Ghost Dance Religion: Poetry Songs and Great Basin Context*. Chicago: University of Illinois Press.

van Schaik, C. P. 1983. "Why Are Diurnal Primates Living in Groups?" *Behaviour* 87:120–44.

van Schaik, C. P., and J. A. R. A. M. van Hooff. 1983. "On the Ultimate Causes of Primate Social Systems." *Behaviour* 85:91–117.

Van Valen, L. 1973. "A New Evolutionary Law." *Evolutionary Theory* 1:1–30.

———. 1976. "Ecological Species, Multispecies, and Oaks." *Taxon* 25 (2–3): 233–39.

Verginelli, F., C. Capelli, V. Coia, M. Musiani, M. Falchetti, L. Ottini, R. Palmirotta, A. Tagliacozzo, I. De Grossi Mazzorin, and R. Mariani-Costantini. 2005. "Mitochondrial DNA from Prehistoric Canids Highlights Relationships between Dogs and South-east European Wolves." *Molecular Biology and Evolution* 22 (12): 2541–51.

Vilà, L., P. Savolainen, J. E. Maldonado, I. R. Amorim, J. E. Rice, R. L. Honeycutt, K. E. Crandall, J. Lundeberg, and R. K. Wayne. 1997. "Multiple and Ancient Origins of the Domestic Dog." *Science* 276:1687–89.

Voilochnikov, A. T., and S. D. Voilochnikov. 1982. *Hunting Laikas* [in Russian]. Moscow: Lesnaya Promyshlennost.

von Franz, M. L. 1995. *Creation Myths*. Rev. ed. Boston: Shambhala Books.

vonHoldt, B., J. P. Pollinger, D. A. Earl, J. C. Knowles, A. R. Boyko, H. Parker, E. Geffen, M. Pilot, W. Jedrzejewski, B. Jedrzejewska, V. Sidorovich, C. Greco, E. Randi, M. Musiani, R. Kays, C. D. Bustamante, E. A. Ostrander, J. Novembre, and R. K. Wayne. 2011. "A Genome-wide Perspective on the Evolutionary History of Enigmatic Wolf-like Canids." *Genome Research* 21:1294–305.

vonHoldt, B. M., J. P. Pollinger, K. E. Lohmueller, E. Han, H. G. Parker, et al. 2010. "Genome-wide SNP and Haplotype Analyses Reveal a Rich History Underlying Dog Domestication." *Nature* 464:898–902.

von Stephanitz, M. 1921. *"Der deutsche Schäferhund in Wort und Bild": Verein für Deutsche Schäferhunde*. https://archive.org/stream/derdeutschescoostep #page/n7/mode/2up.

Voth, H. R. 2008. *The Traditions of the Hopi*. London: Forgotten Books. Originally published in 1905.

Walker, B. L. 2005. *The Lost Wolves of Japan*. Seattle: University of Washington Press.

Wallace, E., and E. A Hoebel. 1948. *Comanches: Lords of the Southern Plain*. Norman: University of Oklahoma Press.

Walters, J., C. A. Annett, and G. Siegwarth. 2000. "Breeding Ecology and Behavior of Ozark Bass *Ambloplites constellatus*." *American Midland Naturalist* 144 (2): 423–27.

Wang, X., and R. H. Tedford. 2008. *Dogs: Their Fossil Record and Evolutionary History*. New York: Columbia University Press.

Watson, L. 2016. "One Gene Doth Not a Hybrid Make—The Clever Semantics of Classification of the Dingo as 'Wild-Dog.'" Western Australian Dingo Association. http://www.wadingo.co7/Hybrid.html.

Wayne, R. K. 2010. "Recent Advances in the Population Genetics of Wolflike Canids." In *The World of Wolves: New Perspectives on Ecology, Behavior, and Management,* edited by M. Musiani, L. Boitani, and P. C. Paquet, 15–38. Calgary: University of Calgary Press.

Wayne, R. K., and S. Jenks. 1991. "Mitochondrial DNA Analysis Implying Extensive Hybridization of the Endangered Red Wolf *Canis rufus.*" *Nature* 351:565–68.

Welsch, R. 1992. *Touching the Fire: Buffalo Dancers, the Sky Bundle and Other Tales.* New York: Villard Books.

White, D. G. 1991. *Myths of the Dog-Man.* Chicago: University of Chicago Press.

Wilson, D. E., and D. M. Reeder. 1993. *Mammalian Species of the World.* 2nd ed. Washington, DC: Smithsonian Institution Press.

Wilson, E. O. 1975. *Sociobiology: The New Synthesis.* Cambridge, MA: Harvard University Press.

———. 1978. *On Human Nature.* Cambridge, MA: Harvard University Press.

———. 1994. *Naturalist.* Washington, DC: Island.

Wilson, P. J., S. Grewal, I. D. Lawford, J. N. M. Heal, A. G. Granacki, D. Pennock, et al. 2000. "DNA Profiles of the Eastern Canadian Wolf and Red Wolf Provide Evidence for a Common Evolution Independent of the Gray Wolf." *Canadian Journal of Zoology* 78:2156–66.

Wilson, P. J., S. Grewal, F. F. Mallory, and B. N. White. 2009. "Genetic Characteristics of Hybrid Wolves across Ontario." *Journal of Heredity* 100: S80–89.

Wilson, P. J., S. Grewal, T. McFadden, R. C. Chambers, and B. N. White. 2001. "Mitochondrial DNA Extracted from Eastern North American Wolves Killed in the 1800's Is Not of Gray Wolf Origin." *Canadian Journal of Zoology* 81:936–40.

Woolpy, J. H., and B. E. Ginsburg. 1967. "Wolf Socialization: A Study of Temperament in a Wild Social Species." *American Zoologist* 7 (2): 357–63.

Worster, D. 1993. *The Wealth of Nature.* Oxford: Oxford University Press.

———. 1994. *Nature's Economy: A History of Ecological Ideas.* 2nd ed. Cambridge: Cambridge University Press.

Wrangham, R. 1980. "An Ecological Model of Female-Bonded Primate Groups." *Behaviour* 75:262–300. doi:10.1163/156853980x00447.

Wrangham, R. W., and D. Peterson. 1996. *Demonic Males: Apes and the Evolution of Human Aggression.* Boston: Houghton Mifflin.

Wroe, S. 1996. "An Investigation of Phylogeny in the Giant Extinct Rat Kangaroo *Ekaltadeta* (Propleopinae, Potoroidae, Marsupialia)." *Journal of Paleontology* 70:681–90.

Wroe, S., J. Brammall, and B. N. Cooke. 1998. "The Skull of *Elkatadeta ima* (Marsuplialia): An Analysis of Some Marsupial Cranial Features and a Reinvestigation of Propleopine Phylogeny, with Notes on the Occurrence of Carnivory in Mammals." *Journal of Paleontology* 72:738–51.

Zeder, M. A. 2006. "Archaeological Approaches to Documenting Animal Domestication." In *Documenting Domestication: New Genetic and Archaeological Paradigms,* edited by M. A. Zeder, D. G. Bradley, E. Emshwiller, and B. D. Smith, 171–80. Berkeley: University of California Press.

INDEX

Page numbers in *italics* refer to illustrations.

Aboriginals, 20–21, 36, 47, 56, 75, 81, 125–42, 178, 280–81

Adirondacks, 167

Afghan, 46, 126

Africa, 5, 43, 56, 63, 73–74

African wild dog (*Lycaon pictus*), 58–59, 60, 79

Ainu, 108, 121, 122–23, 134, 139, 158, 280, 281

Akita, 46, 121, 122, 126, 138, 285

Alaska, 86, 94, 141, 155, 223, 282–83

Alaskan barren ground wolf (*Canis lupus tundrarum*), 252

Alaskan husky, 44, 45

Alaskan malamute, 11, 25, 205, 211, 212, 214, 225

Allen, Durward, 57

alpha female, 1, 4, 72, 137, 162

Alsatian. *See* German shepherd

Altai Mountains, 46, 96, 108–9, 111, 113

amensalism, 36

American Kennel Club (AKC), 22, 46, 95, 205, 207, 225

American Society of Mammalogists, 24–25

American Veterinary Medical Association (AVMA), 218–19, 225, 233, 234

Anderson, Eugene, 9–10

Angara River, 111

Animal Connection, The (Shipman), xiii

animal control, 27, 205, 212, 215, 236, 243, 245–46

Anishinaabe, 81

Annett, Cynthia, 208, 255, 256, 258–59, 260–62, 266, 269

ants, 192

apes, 64, 67–68, 73, 76

Arapaho, 145

Archaea, 64

archeology, 83–104, 193

Arikara, 145, 152

Asian elephant (*Elphas indicus*), 200

Assiniboine, 145

Audubon, John James, 223

Augustine, Saint, Bishop of Hippo, 188

Australia, 5, 9, 15, 20, 36, 74, 97, 107, 125–42, 216. *See also* Aboriginals; dingo